PLUGGED IN

.

PLUGGED IN

HOW AI WILL ENHANCE AN EVOLVING SPORTS WORLD

GREG WILLS

NEW DEGREE PRESS

PLUGGED IN

How AI will Enhance an Evolving Sports World

ISBN   978-1-64137-062-2 *Paperback*

            978-1-64137-063-9 *Ebook*

This is dedicated to my family, including Ryan,
who is watching out for me always.

CONTENTS

INTRODUCTION 9

WHAT IS IT ALL ABOUT? 27

FOR THE LOVE OF THE GAME(S) 41

PREDICTIVE STATISTICAL ANALYSIS 57

COMPUTER VISION 75

SPORTS BETTING 93

DATA COLLECTION USING WEARABLES 105

TRAINING, RECOVERY, AND INJURY PREVENTION...........115

FAN ENGAGEMENT... 127

BEING A FAN.. 139

WOULD WE EVER REPLACE THE HUMAN ATHLETES? 155

REFERENCES.. 167

ACKNOWLEDGEMENTS...................................... 175

INTRODUCTION

——

September 4, 2002

Up 11-0 against the Kansas City Royals, the Oakland Athlet-ics were on top of the world. Coming off winning 19 games straight, the Billy Beane-led Athletics looked to extend their streak to 20 games, then an American League record. With a low budget and major free agency losses, the Athletics were projected to go nowhere at the beginning of the season. Now here they are with an 11 run lead in what would be their 20th straight win.

But slowly the Athletics (A's) lost their lead, having given up 10 runs going into the 9th inning. With the crowd roaring at his back, A's pitcher Billy Koch surrendered a run scoring single, tying the game up at 11. Watching a seemingly insurmountable lead demolished, the fans deflated like a balloon. The record was almost a guarantee after the A's put up 11, right?

Enter Scott Hatteberg, the first baseman for the A's. Hatteberg, who thought he was done with baseball before the 2002 season after a severe injury, steps up to the plate and blasts a moonball over the right-center fence in the Oakland Alameda Stadium, sending the Athletics to their record-breaking 20th win.

A popular story, no doubt. A vast majority of sports fans know this story from *Moneyball*. And if you've read the book or seen the movie, you know the story of Billy Beane, the GM of the A's at the time. By focusing on obscure performance statistics, the Athletics crafted their lineup with undervalued players that they could afford. This streak marks the advent of advanced sports analytics in analyzing player talent in the game of baseball. In just 15 years, sports analytics has grown to be one of the most important measures for evaluating players in all sports, getting more advanced by the day.

The necessity for teams to compete with the Athletics drove a revolution in baseball.

A similar revolution is happening across all professional sports right now. But it is not based solely on statistical analysis of players.

This revolution is driven by *Artificial Intelligence.*

* * *

While the statistical revolution in baseball was a huge transformation for the sport, artificial intelligence has the power to make a bigger impact on the sports industry. The transformation, ushered in by the 1990s Oakland Athletics, has been expanded by savvy executives such as Theo Epstein and Dave Dombrowski. Due to the sheer power that AI possesses, AI will lead to a bigger transformation in sports than the Advanced Analytics movement seen in *Moneyball.*

Welcome to the new revolution.

The advanced analytics movement, while powerful, only assisted teams in assessing talent. AI reaches not only the advanced analytics area, but also several other areas that have a great impact on all sports. The areas that will be most affected by AI are detailed in each separate chapter:

- Predictive Analytics
- Sports Gambling Industry
- Training and Injury Prevention
- Data gathered from wearable technology
- Sports Journalism
- Fan Engagement

As these industries become progressively more affected by AI, their specific impact will become clearer. The outcome of the impact of artificial intelligence remains to be seen. With such a wide reach, AI has the ability to affect these areas so significantly that it will change the way teams operate and how fans consume sports. It is so wide reaching that the replacement of athletes altogether could become a possibility. Going beyond Billy Beane and the analytics movement in baseball, AI will have a greater effect on the sports industry than we can imagine.

Across the board Artificial Intelligence seems to be a new state of affairs in our modern society. Seemingly every day a new article is published about new AI technologies present in all sectors

- AI is projected to replace 16% of Americans jobs by 2020.
- By 2025, the AI market is projected to surpass the $100 billion mark.

The sports industry is a similar industry that has been growing in popularity over the years. There are many applications of AI in sports already, and as AI makes advances, so will the profitability of the professional sports industry.

The North American sports industry alone is projected to hit $73.5 billion in revenues by 2019.

- Lebron James, arguably the best player in the NBA, is set to make nearly $31 million in salary alone for the year 2018.

In a year that Lebron James makes that much in salary alone, the profitability of the sports industry cannot be denied.

As an aspiring computer scientist, I have always been fascinated by Artificial Intelligence and the endless possibilities it presents. As a freshman in college, I enrolled in an AI seminar that covered a wide range of topics from neuroscience to philosophy.

I vividly remember watching a video that emerged during the course which our professor showed to us. In an interview, Elon Musk is quoted as saying, *"We are summoning the demon with AI."*

While my research has not shown this to be true yet, I know it is certainly a possibility. AI has proven itself to solve very narrow tasks designated by the programmer. This notion only encourages my interest in AI in application to the sports world. What effects will AI have on the sports that I have grown to love? How will it affect my involvement as a fan? The answers to these questions are changing every day, yet I believe it will have a profound effect on the lives of everyone across the world.

Here are a few previews of the insights you will gain from the book. Inside you'll learn . . .

- How AI is used in sports today
- How AI has grown to be a big part of the sports industry
- How different areas are affected by new technologies
- The possible limitations of AI in sports
- The implication of constantly developing AI technologies
- The fundamental transformation of the sports industry
- How to be successful in an evolving industry

* * *

It is hard to get a comprehensive understanding of AI in the sports world. I do not believe AI has had a profound enough effect on the sports industry yet to gauge the change it might bring about. Looking at the separate markets, it is easy to imagine the possibility for growth in the intersection. While not widely acknowledged yet, AI already is having a profound effect on many sports that we know and love.

To the normal fan, it is hard to distinguish between new technology and AI specific technology. It is important for you to understand that AI has been around for nearly a century, and it will continue to grow rapidly with the existence of modern computing systems. The integration of AI into sports is a reality that needs to be faced and addressed by all. The improvements that can be made to the overall quality of each athlete and their athletic performance is drastic. New technologies have been helping athletes improve throughout the history of sports, and AI is the next big technology to make these advancements.

AI in our lives. For those of you who chose to pick up this book and read it, thank you. You have shown that you believe in me and that I can provide you with the tools you need to go forward in this ever advancing world that we call Artificial Intelligence. After doing much research, I have gained much knowledge on the development of AI and its applications across all sectors of our life. For a guy who loves sports, it was only logical to write about AI in sports.

Artificial Intelligence is everywhere.

The idea of writing about sports was a natural conclusion because I follow sports obsessively. Although people may not realize it, AI is everywhere in sports. I guarantee that you will start to hear more about the integration of sports and AI. As it has done over the past decade, AI has become a huge part of our lives. AI will fundamentally change the sporting industry, and in a greater way that we can imagine.

Take Siri for example. This conversational bot is the product of AI and is in every Apple iPhone in the world. Think about how many of your friends have iPhones, or even yourself. For someone who only uses Siri for the weather and the occasional Google search, I was amazed at how advanced AI has become just during the few months I have been doing research.

Siri has little to do with sports, but provides a good template for how quickly artificial intelligence can improve.

I will preview some stories that you will encounter in the chapters that exemplify AI's current involvement in sports.

- **Sailing**: GPS tracker on boating equipment provides real time and accurate recordings for race results, speeds, and tactical decision making.
- **Fencing**: Sensors in the gear allow the judges and viewers to see the exact moment and location of a hit
- **Tennis**: An arena's screen will display the speed of a hit in real time.
- **Hockey, tennis, rugby, gymnastics**: Advanced camera vision technology assists referees in making decisions to eliminate controversy.
- **Basketball**: Computer vision technology is being used to enhance a viewer's experience by displaying advanced statistics in real time.
- **Soccer**: Neural networks (AI based computer systems modeled on the human brain) are being used to determine outcomes of matches based on advanced statistics gathered by cameras.

- **Horse racing**: AI technology is used to attempt to determine the superfecta at the Kentucky Derby.
- In nearly every sport, you can find a way in which artificial intelligence is being used to enhance performance or enhance viewer experience.

But sometimes people fail to understand the difference between Artificial Intelligence and just good programming. Artificial intelligence is the theory of computer systems that are able to perform tasks that normally require human intelligence. AI gives us the ability to complete tasks quicker and more compactly than otherwise possible. Distinguishing the difference between these theories requires some thought and research, but as you read, it will become clear what AI is, how it's used in sports, how it can help us, and how it will affect the sports industry forever.

* * *

Why Sports and AI. As a kid I always knew I wanted to be a professional baseball player. As kids, we have a naïve ambition that one day we might be the President of the United States of America or a Major League Baseball player. Don't get me wrong, I truly believe that following your dreams is an admirable ambition, but as I grew older, my athletic ability was a big indicator that I was not going to be an MLB player. While I had the passion, I certainly lacked the God-given athletic ability to keep advancing my baseball career.

Given that I was not born to be a star in the majors, I had to focus my interests elsewhere. From the start I wasn't drawn to technology or programming as I was to baseball. The development of my love for computer science was a slower process. I knew I didn't want to be a humanities major as I didn't love writing essays. It's ironic now that I have written a book, nearly 25,000 words . . . If I wasn't going to be an English or History guy, I knew I had to be either a Math or Science guy. The decision to major in computer science developed out of a desire for a job out of college. I got to where I am today academically by a realistic view of my opportunities. With this book though, I have learned so much on the topic, and I want to help others understand its importance.

Part of the reason I have decided to write this book is not only my love for sports and technology, but also AI's ability to truly help us improve all facets of the sports industry. From player development, statistical analysis, and injury prevention to streamlined business strategies, different applications of Artificial Intelligence will go a long way in improving the overall experience of sports for people across the globe.

* * *

Misconceptions. My research has taught me so much about

the rise of AI in general. From doing interviews with professors experienced in the field of AI, I was able to construct an accurate landscape of AI now and how it has progressed to this point. AI has evolved and grown so much in the past few decades. It seems as though the media has popularized the rise of Artificial Intelligence with stories on Elon Musk and other creators.

AI in sports has increased as well. As AI technologies get more popularized, they are incorporated more intricately into the professional sports we know and love.

Another popularized notion on AI is that it will "take over the world" like in the movies *Terminator* or *Ex Machina*. I have found that we are not nearly close enough for this to be the case. AI has proven itself to handle simple, narrow tasks dictated to it by humans. It can learn through repetition, but still, to this point, there is no threat of humans losing control. So, if AI can only be used to complete narrow tasks, then how helpful is it?

"By far, the greatest danger of Artificial Intelligence is that people conclude too early that they understand it."—Eliezer Yudkowsky

I will never claim that I am an expert in artificial intelligence. Although I have done the research on sports specific Artificial Intelligence that goes beyond the knowledge of most people,

I think the beauty of AI is that not many people are able to fully grasp its utility.

I want to clarify that the AI I will cover is one that may not be your idea of AI. I am not talking about artificially intelligent beings, but more of the techniques used to produce systems that can compute more than can the human mind.

You will learn what exactly these techniques and categories of AI are and how they have been and will be used to affect the sports industry in a significant way.

If we are to use these techniques in a narrow but effective way, we will have success in improving the profitability of the sports industry. Ultimately, AI will change the way fans view sports, the way general managers select their players, the way athletic trainers help their players avoid injuries, and the way players develop their skills. It will greatly affect statistical models of analysis and already does to this day. The ability to analyze players' motions through AI improves the categories of statistics on which we judge players' performance.

AI, if employed and used correctly, has the ability to affect every aspect of every sports that exists today. The common misconception that AI is something to be afraid of can be tossed aside. Artificial intelligence will grow to be a billion dollar market soon.

The use of AI is undervalued and this book shows how the sports industry is being transformed.

* * *

Necessity for Transformation. This book is going to describe how AI is currently involved in sports and how I see it applying in the future. My view is that AI certainly does have limitations, but as we increasingly use AI in our day-to-day lives, we will understand those limitations more clearly. By knowing the limits, we will know exactly how AI can be applied in each facet of sports.

You can begin to form your own predictions of how AI will change the sports industry in the future. This is by no means limited to those involved in the sports industry. The average sports fan can predict how their viewing experience will be enhanced, or how they will be better able to analyze their favorite players' tendencies.

Mainly, though, this book is aimed at people involved in the sports industry. You will hear from major contributors to the AI movement in sports and major advancements made by these experts. In an industry where only a slight advantage is needed to be a successful organization, AI can make the difference. But in order to use AI to improve an organization, one needs to understand the current landscape of the sports industry in relation to artificially intelligent programs and systems.

The combination of the sports and AI industries can be very profitable. Professional sports is a multi-billion dollar industry that is growing by the day. The sports industry is already in the process of being transformed by AI, and it will be of the utmost importance to understand it going forward.

From general managers and owners, to media giants, AI will change the way in which their organizations function.

* * *

Through the stories that you will encounter in this book, it will become clear how to maximize success using AI in the sports industry. AI has already proven to be incredibly useful in our society today. Just the idea of self-driving cars, while still in progress, can help humans maximize productivity. Our brains are powerful, yet AI can go significantly beyond our human capabilities and limitations.

With both industries being highly profitable, it is only logical that the combination could prove to enhance both industries' capabilities.

And by improving the product of professional sports, we also increase fan engagement. I believe that AI can combine and maximize two extremely large and profitable industries, and transform them into one giant industry.

My insights come from a place of amazement and wonder at what is possible through combining these industries. Previously, I was aware that my fanaticism for most professional sports was higher than most. But after doing my research, I am even more excited about the advances to come. AI is only starting to dig its claws into the sports industry and vice versa. If it can affect the sports industry now, imagine what it can do 10 years from now. And by improving these technologies, they will get less expensive to manage and cut costs across the board.

I think my fanaticism for sports and love for technology further proves the relevance this topic has today. In a world where it is a rare night that you cannot find a professional sports game on television, sports rules. Improving a product that people already consume on a regular basis in a significant way can further the sports industries' dominance.

"This is a man who was 23 years old when he theorized the idea of creating a programmable machine, and in that way, Turing foresaw computers and artificial intelligence. These were revolutionary ideas at that time."—Morten Tyldum

Speaking about Alan Turing, one of the fathers of Artificial Intelligence, Tyldum speaks about his revolutionary ideas. What were once revolutionary ideas have become somewhat commonplace in this day and age. By reading this book, you will learn how to revolutionize the sports industry, using ideas being improved upon since the days of Turing. And sports have been constantly improving over the years just like AI, turning themselves into billion-dollar industries. New advancements are coming every day, but AI technologies will grow in sports and ultimately change the sports landscape that we know forever.

It is important to understand that AI has been around nearly a century, and it will continue to grow rapidly with the existence of modern computing systems. AI is a necessity in our modern society.

In order for the sports industry to make advancements, AI needs to be integrated into sports. Like the advanced analytics movement, new technologies constantly have made huge impacts on the sports industry.

Understanding how to incorporate AI in sports will only lead to a more successful and profitable industry. The inherent power of both industries is clear. The combination will serve to advance sports beyond what anyone could've imagined in recent history.

Let's go.

CHAPTER 1

WHAT IS IT ALL ABOUT?

———

Artificial Intelligence, at this point in time, will not flip a switch and "take over" as we see in the movies. While AI technologies have made gigantic strides over the years, we still have a long way to go before we need Arnold Schwarzenegger to come save us from the big, bad SkyNet.

In a recent experiment, Facebook had to shut down its AI project because the machines began "talking to each other" in a language that they seem to have created themselves. While this can mean many different things, those without proper knowledge of the project can only be left pondering the scary implications of the rise of AI. It may not be as scary as we think.

That being said, the applications of AI in sports are endless.

* * *

Artificial Intelligence is a complex topic to define and there are many systems that lack a black and white line distinguishing it from regular computer science.

One such definition of Artificial Intelligence comes from the Merriam Webster dictionary: "a branch of computer science dealing with the simulation of intelligent behavior in computers, and the capability of a machine to imitate intelligent human behavior."

Techopedia defines it as . . .

"a branch of computer science that aims to create intelligent machines."

Nils John Nillson, 1998 . . .

"Artificial intelligence, broadly (and somewhat circularly) defined, is concerned with intelligent behavior in artifacts. Intelligent behavior, in turn, involves perception, reasoning, learning, communicating, and acting in complex environments.

These definitions describe a high-level delve into Artificial Intelligence and its goals for recreating human functions. There are many theories that take stances on what actually constitutes an artificially intelligent being. This is the classic view of AI that most people hold, probably derived from movies like *Terminator* or *Ex Machina*.

"What we have are limited targeted use cases for AI. . . . What we don't have is what's called artificial general intelligence."— **Oren Etzioni**

One clarification is that I will not talk much, if at all, about creating intelligent beings. It is supposed that these beings could handle any reasonable task able to be performed by humans. Rather, I will discuss tactics used to create artificially intelligent information to be used in other ways to facilitate daily tasks in a much more applied fashion.

But people use AI in their everyday lives more often than one might think. Products like Siri, or even a simple Google search use artificial intelligence to enhance their products.

Using anonymous location data from smartphones, Google Maps can analyze the speed of movement of traffic at any given time. Its calibration of traffic allows its AI to suggest the fastest route to work for the daily commuter. In the future, AI will shorten your commute with self-driving cars, reducing accidents by up to 90%. Ride sharing apps like Uber will reduce the number of cars on the road by 75% as they get increasingly more efficient.

The timeline for these changes is unclear, but it is coming soon. The journal Business Insider Intelligence predicts that self-driving cars will be available in 2019.

"It's a years thing, not a decades thing."—Travis Kalanick, Uber CEO

Another great example of AI in use today is fraud protection for credit cards. Most, if not all, banks use artificial intelligence to detect fraudulent credit card purchases. Their computer system is given a very large sample of fraudulent and non-fraudulent purchases. It is then asked to look for signs that a transaction falls into either category. After enough training, the system is able to spot whether a transaction is fraudulent or not, allowing the bank to freeze your account to prevent more fraudulent transactions.

Humans cannot review all daily transactions in an efficient amount of time, but the computer can through artificial intelligence. FICO, the company that determines credit scores, uses neural networks, a form of AI, to predict fraudulent transactions. Many factors go into the prediction including recent frequency of transactions, transaction size, and the kind of retailer involved.

I expect that our investing strategies in the future will be largely determined by AI. Companies, such as Betterment and Wealthfront, are attempting to automate the best practices of seasoned investors in their systems. By offering lower cost advice, these companies could rise to prominence in the investing world. In 2016, Wealthfront announced it was taking an AI-first approach, *"an advice engine rooted in artificial intelligence and modern APIs, an engine that we believe will deliver more relevant and personalized advice than ever before."*

Even targeted advertising, a constant in our daily lives, uses AI to predict what a consumer might like. A scary trend is arising in targeted advertising where our phones and computers track what we search, say, and do. While it seems to be harmless now, the security and safety of AI technologies is something to watch as we move forward in the development of AI.

The importance of these examples is that they are constantly evolving as AI technologies get more advanced. AI will become the norm for companies looking to get ahead of their competition. As AI gets more advanced, I would expect it to become more accurate with its predictions. Predictive AI is on the rise, and will take center stage as the years go on.

* * *

As you can see, artificial intelligence encompasses a lot more than just mimicking human behavior with, say, robots. Two highly useful techniques for creating AI technology are machine learning and deep learning.

Machine learning (ML), as defined by the journal Expert System, is *"an application of artificial intelligence (AI) that provides systems the ability to automatically learn and improve from experience without being explicitly programmed. Machine learning focuses on the development of computer programs that can access data and use it learn for themselves."*

Mark Maloof, a Director of Undergraduate Studies and a professor in the department of computer science at Georgetown, defines machine learning as *"given data, develop or use computational methods to build models that, one, predict something about new data, and two, provide a better understanding of the data itself."*

Essentially the primary aim is to allow computer programs to learn automatically without human intervention and adjust their actions accordingly, beyond its human programming.

We can break machine learning (ML) up into 3 components.

Representation language. This is the language used to build ML models

Training algorithm. This uses training examples to create models. Basically, it uses large sets of data to create models. It generates clauses consistent with the data. It then finds maximum likelihood estimates for prior statistical distributions.

Prediction algorithm. This uses the training model to output a prediction for an observation. It uses logical deduction to make a prediction, sometimes using Bayesian theory to make the most probable prediction. Bayes theory describes the probability of an event based on prior knowledge of conditions that may be related to the event.

As these components become more accurate, so will our predictions. And they will get more accurate.

This is the basis of most artificially intelligent systems currently and can be applied to most topics I will cover in this book. Machine learning, given that it is able to analyze data with little human intervention, relies on mass amounts of data to be able to draw conclusions with the data at hand.

Deep learning is a subset of machine learning. It is inspired by our biological understanding of the human brain. Part of this deep learning involves neural networks which, in a sense, attempt to mimic the synapses in the brain and attempt to simulate the way in which humans form connections and in a word, think.

How does deep learning work? I'm glad you asked.

Deep learning boils down to feeding large data sets into what is called neural networks. These neural networks are comprised of logical constructions which ask binary true/false questions, or extract a numerical value. From there, this result is classified based on the data that was passed through. I will talk more about neural networks later. However, I won't make much of a distinction between machine learning and deep learning as the book continues on, as the difference is more technical than I would like to get. It remains important to know the distinction and the background of each application of these AI techniques.

Professor Jeremy Bolton, assistant teaching professor in the Department of Computer Science at Georgetown, described to me the rise and fall and rise again of deep learning. This discussion appears in a later chapter.

* * *

After taking a seminar in Artificial Intelligence in my freshman year of college, I knew that AI was one of my passions. Not only was I extremely interested by the class, I also was intrigued by the future applications of such a powerful and potentially dangerous field. Now I have to admit I was worried about transitioning my passion into enough passion to write a whole book on it. I have learned that finding people to speak with about my book has been more difficult than I anticipated. Finding a balance between sports and Artificial Intelligence was also a struggle, but I believe I was able to strike a nice balance.

My interest in Artificial Intelligence began back even before college, in the summer before my first year in college. Applying to colleges was one of the most painstaking and stressful experiences of my senior year of high school and my physics grade reflected this experience. After thinking I was done with college applications forever, an interesting possibility came across my desk. Taking time from my "busy" summer lifeguarding schedule, I took the time to read through a pamphlet I received. It included many different freshman seminars including one on Artificial Intelligence, named Artificial Intelligence: From NAND to Consciousness. Luckily enough I sat down and applied almost immediately, and to this day I know I made a choice that would affect my college experience forever.

I was lucky enough to be accepted into this seminar and planned to take this course in the fall. The professor, Mark Maloof, turned out to be one of the best professors I've had here at Georgetown and to this day is still my advisor.

The course dealt with the ethics and evolution of AI over the years and as the course moved along, we learned about the current applications of AI. From self-driving cars to neural networks, I was enthralled with the profound effect these innovations can and do have on our society today.

Now, as a junior, I decided to sit down with Professor Maloof and continue our discussion of AI dating back to two years prior. We delved into his expert opinions on machine learning and the evolution of AI through the years.

He, too, was inspired to pursue AI after developing an AI project at an internship during his college years. Years later, after he earned a PhD in Machine Learning at the University of Georgia, he taught his first class in AI at Georgetown University. Immediately he knew it was his calling.

In 2005, Professor Maloof took a sabbatical and worked at the MITRE Corporation on a project aimed at dealing with insider threats. He remarked on the loads of real data that they dealt with—an important aspect to the development of machine learning. Throughout my research on Artificial Intelligence, this fact has caught my eye the most. The necessity for humungous amounts of data is a prevailing theme in successful AI projects. As the amount of data all around the world gets progressively larger, it isn't hard to imagine that AI will only continue to grow. As a result of his team's work, the program was not only used operationally, it gained a patent and even won an industry award. Mark and his team not only built a great, highly operational project, but it was used by the MITRE Corporation as one of their products. After I pressed him more about his project, he admitted he could not say more as it was classified.

Another interesting takeaway from my interview with Professor Maloof was that although AI technologies seemed to be captivating the attention of the media recently, AI development has been going on for years, far before the media was ready to report on it.

For example, self-driving cars, while they are on the verge of becoming popularized, are no new phenomena. Self-driving cars have been a thing for decades, and specifically in 1996 when a car drove from Pittsburgh to San Diego on full automation. Given the recent advent of Google's and Tesla's automated cars, you might think this is brand new technology. Quite the contrary, actually.

Now, where will AI progress to in the future? Who's to say? But Professor Maloof gave good insights on how it will continue to progress. While we seem to be making leaps and bounds of progress, we are really incrementally developing algorithms and techniques that make AI more accessible to the general public.

In Professor Maloof's mind, a combination of 3 things stands out in how AI has become more prevalent in our daily lives.

The first is the size of computer networks and their processing power. Google's data centers are able to handle and process approximately 1.2 trillion searches per year. To put this into perspective, not one of Google's data centers is in the top 10 of data centers worldwide. These centers range from 400,000 to 1.1 million square feet. The average US house is only about 2700 square feet. About 400 average sized US homes can fit inside these data centers. Let that sink in. An insane amount of data has to be processed on the internet, and has only grown larger over the years.

Second, computer networks combined with easy to use software have contributed to the growth of AI as well. To use the example of Google again, you can simply type a query into Google and your results are pulled up almost immediately. Simply typing a phrase into Google gives you every answer you could possibly want.

The third is refined algorithms. Without a background in computer science it is hard to understand the importance of algorithms in the effective execution of programs. Say you are writing a program and have to sort numbers in a list. You can write one algorithm that will sort your numbers in, say 5, minutes. While this doesn't seem terrible, consider your Google search. It pulls results in under a second. Now we can compare the same program, but written more efficiently it can sort your list of numbers in a millisecond. The revision and incremental refining of these algorithms can only serve to give the public easier access to large amounts of data.

The combination of computer networks, the progress in easy-to- use software, and refined algorithms has created this advancement and publicity for AI projects that did not see the light of day before.

Finally, we got around to the topic of sports, the topic you've come here to read about. Professor Maloof suggests there will come a point in time when we will have to decide to phase out analytics in sports and convert to software run by AI. The precedent has already been set that every known game out there, such as poker or chess, has an AI software that can beat any human in existence. While Mark was in college, he noted AI chess bots were just starting to beat chess grand masters at their own game. Now, it is hard to mention a game that cannot be played expertly by an AI bot.

A concession that I will make is that it is hard to replicate human interactions and reasoning, especially in such a physical endeavor as sports. But it is not unreasonable to think that one day no human will be able to beat an AI software in sports as well, no matter the physical application.

CHAPTER 2

FOR THE LOVE OF THE GAME(S)

———

WHY (SOME) ASPECTS OF SPORT WILL LIKELY ALWAYS BE OFF-LIMITS.

"It is not the critic who counts; not the man who points out how the strong man stumbles, or where the doer of deeds could have done them better. The credit belongs to the man who is actually in the arena, whose face is marred by dust and sweat and blood; who strives valiantly; who errs, who comes short again and again, because there is no effort without error and shortcoming; but who does actually strive to do the deeds; who knows great enthusiasms, the great devotions; who spends himself in a worthy cause; who at the best knows in the end the triumph of high achievement, and who at the worst, if he fails,

at least fails while daring greatly, so that his place shall never be with those cold and timid souls who neither know victory nor defeat."—Teddy Roosevelt

In 1910, Teddy Roosevelt gave his famous The Man in the Arena line in his speech "Citizenship in a Republic" delivered at the Sorbonne, in Paris, France.

He offers a statement that should help us draw some lines about technology and sport:

The man who is actually in the arena, whose face is marred by dust and sweat and blood; who strives valiantly; who errs, who comes short again and again.

While many things could actually be at risk with the dawn of artificial intelligence, the human competition of sports seems unlikely to be one of the earliest things to go. Inherent in sport is competition. Inherent in competition are human beings. The reason we love sports is not only the competition, but also the displays of humanity across all sports.

One story that comes immediately to mind is Jesse Owens in the 1936 Berlin Olympics. Owens, an African-American track athlete, gathered 3 world records and 4 Gold medals during the Berlin Olympics. Adolph Hitler expected that his German team would achieve great success, led by his belief in an Aryan race. But Hitler refused to acknowledge Owens' significant accomplishments in the face of racism and heavy adversity. Coming during the Great Depression and the rise of Adolph Hitler, Jesse Owens gave Americans a hope for a better world, transcending sports entirely.

The famous picture of Owens at the top of the podium, surrounded by Germans *heiling* Hitler is a powerful moment not just in sports, but in world history. The human element keeps people coming back to watch the sports that they love, and will not be easily eliminated as artificial intelligence continues on its upward trend.

And it's why I'll spend little time—except for some thoughts in the final chapter—about the impact of artificial intelligence on the players participating on the field in the arena.

* * *

The special bond between humans, competition, and sport became more evident during my internship with the Baltimore Orioles.

My high school baseball coach was a former MLB player for the Orioles and put me in contact with someone in the Orioles Alumni department. After my senior year, I participated in a mandatory two week internship program. I decided I wanted to do something that I am passionate about, namely sports.

So, on my first day I showed up at the iconic warehouse at Camden Yards, and I remember getting this special feeling of mesmerization that I actually am working for the Orioles. I wasn't just showing up for a game as a fan, I was going to be a (temporary) employee for the Orioles. Lucky for me the Orioles were on a long home stand for most of my internship. My job was to guide former Orioles players around to the suites so that they could chat and sign autographs with the fans. It was truly a special experience meeting the players, some of whom I looked up to during their playing days. I met players such as Al Bumbry, Larry Bigbie, Chris Hoiles, and Tippy Martinez.

The most memorable conversation I had was with Al Bumbry. Al actually took the time to get to know me and ask about my personal life. He was so outgoing and had a positive out-look on life. Having an Orioles great who I looked up to take interest in my life was something I'll always remember. I not only got to live part of my dream, I also gained valuable experience networking and making conversation with people I was intimidated by.

An inherent part of our culture is change.

Specifically, advancements in technology have been occurring since the dawn of time. Even more specifically, sports have been evolving since their development in the past decades. This is no more evident in any sport than baseball. While an intern for the Baltimore Orioles organization, my hometown team, I worked as an intern for the Alumni Department. My boss, Bill Stetka, is the founder and director of this department and I got to know him rather well over the mere two weeks I shadowed him.

Not only did I meet many alumni, but Bill is also good friends with the current players. He explained to me that over his many years with the Orioles, the biggest thing he noticed was the difference in attitudes towards the game.

A Changing Environment. One particular experience he remembers well was an exchange with Earl Weaver. Earl Weaver has since passed, but at the time he was the charismatic manager for the Baltimore Orioles. Weaver was known as a big believer in the 3-run home run. He relied on it as a means of scoring runs, among other things of course. Weaver reportedly remarked on the quickly evolving landscape of baseball.

The game has changed so much since those days of simple statistics such as batting average and ERA (earned runs against). With the advent of sabermetrics, relying on the 3-run home run was a relic of the past. Bill continued to talk about how Weaver said that he could not have predicted the way the game would change in just 40 years. From 3-run home runs to advanced analytics, the game of baseball has changed forever. Given this change, I can guarantee that the game of baseball will change drastically over the next 40 years as well.

The driving force behind the change will be AI. There are already so many AI products that are helping to advance all sports around the globe. From big data analytics, to computer vision, to machine learning, to neural networks, to GPS tracking, to whatever else AI has in store, it will have a lasting impact on the sports we know and love today.

A Community. This being said, in my short time with the Orioles, my biggest takeaway was not only one of awe for the inner workings of the Orioles operations, but also my admiration for the Orioles community that exists beyond the ballpark. I was lucky enough to see not only the former players' perspective and their relationship with Bill and the alumni department, but also the fans utter admiration for these players they considered their heroes when they were growing up. The Orioles alumni loved coming and giving back to the community that provided them with so much in their playing days. And the fans loved interacting and getting autographs from players they adored throughout their career.

This sense of reciprocal respect is what makes sports what they are today. Without the fans, the players cannot play, and without players, the fans have nothing to root for. The sense of community I felt in my time with the Orioles shed a light on the personal side of sports that sometimes goes underappreciated.

Given that AI involvement is trending upwards, going forwards it will be a tough balance to strike between technology and human beings. There has to be a limit to the expansion of AI, but one that we have not found yet. If we were to lose the sense of community and personal touch in sports, we might lose the reason we love sports in the first place.

* * *

Having played baseball all my life, I have grown fond of the game at every level. From t-ball to Major League Baseball, I can say I enjoy every level of the sport. Well, watching my brother Matt's t-ball games was hard to enjoy, but you get the point. My love of the game thus far has been that of a fan, but I have always had an interest in getting more involved in the sports world.

Realizing my strengths as a computer scientist, I knew I would only be able to write a book on topics I could take an interest in.

In my interview with Pat O'Conner, the President of Minor League Baseball, I was nervous! What do I say to someone who has had such a profound impact on the game that I love?

Immediately Mr. O'Conner soothed my worries as he was a great guy who had a lot to say. This was one of my best interviews and has shaped my idea for my book immensely. He has an impressive track record in his almost 10 years as the president, but has had 34 years of experience in professional baseball.

His willingness to share his stories says more about him than his period of unprecedented leadership of Minor League Baseball (MiLB). He provided an optimistic but realistic approach to the applications of Artificial Intelligence in professional sports. This realistic approach brought my imagination down to earth a little bit.

Presiding over 160 minor league teams, Mr. O'Conner has dealt with every aspect of the game, but deals mostly with the business side of baseball. A few of his insights stuck out to me.

A Quick Progression. He talked about the progression of technology over his 34 years in professional baseball. He remembers looking at handwritten records of the sale of hot dogs for a team to determine how many hot dogs to order for a given period of time. Now nearly all stadiums use algorithms to determine how many hot dogs to order based on decades of stadium data, which trains the algorithm to know what to do. This process of training the algorithm is called machine learning, a subtle application of artificial intelligence which has worked its way into the minor leagues.

Stadium 1, a service that does exactly as described above and more, utilizes a stadium's resources more efficiently. It is even able to tell you where to move hot dogs during the game in order to maximize sales and get the most out of your inventory. It can also tell you when to let employees go home if a stand hasn't made a sale in a certain number of minutes. These are just a few examples of what Stadium 1 is capable of through machine learning.

This is one small example of technology beginning to infiltrate our daily lives based on affordability. As time goes on, certain technologies get less expensive as processes are made more efficient. While efficiently managing a stadium leads to more revenue, it is important to strike a balance between technology and human interaction.

While numbers tell a lot of the story, it is important to remember the human element when making decisions. Mr. O'Conner remarked that the most successful organizations are able to find a happy medium between technology and people when determining players to go get.

The Human Element. An important anecdote from Mr.

O'Conner resulted from his story on negotiating deals with Major League Baseball concerning the Professional Baseball Agreements. It provides a good example of the importance of human interaction within the professional sports industry.

For those of you who don't know, the Professional Baseball Agreements define the relationship between MLB and MiLB and have framed the modern-era structure for Minor League Baseball while creating a healthy financial environment for all clubs in the association.

Mr. O'Conner inherited a situation of bad negotiations between leagues in 1990. After narrowly avoiding a split with the MLB, Mr. O'Conner tirelessly worked to change the attitude between leagues after deciding that they should never get to the point of almost splitting again. After years of healing the relationship between leagues, they can boast of nearly 30 years of healthy negotiations and successful Professional Baseball Agreements. This type of culture change and negotiation cannot be brought about by technology, but only through creating and maintaining healthy human interactions.

One thing that Mr. O'Conner helped organize in these agreements was the packaging of digital rights for Minor League Clubs. In coordination with Major League Baseball Advanced Media (MLBAM), he ensured that MiLB is on the forefront of technology. I will talk about the importance of MLBAM in a later chapter.

This agreement allows fans to gain access to the clubs and the players through websites and technology, which helps to create growth of fan engagement which leads to monetization. One of his best insights, and I think what drives innovation of technologies such as AI, was that new money is better than more money. I'll explain.

While getting more money is always good, innovation of new products and technologies only serves to develop a business and grow it. Being able to monetize innovation is hugely important in creating fan engagement and therefore leads to more money than before.

He also brought some realism toward the end of the conversation. While "anything is possible," the reality of increasingly advanced technology comes with the possibility of quickly obsolete technology. One thing he analyzes when authorizing new technologies is the build-out time, cost, and return on investment. These are all realities that could limit technological innovation. One issue with building, developing, and testing a product is its possibility to be obsolete by the time it is ready for use.

Mr. O'Conner has played a huge role in the evolution of the Minor Leagues and his experiences are indicative of changing technology development over the years. MLBAM and Stadium 1 are great examples of AI being used in professional sports in areas that you might not suspect. While these are revolutionary technologies, going forward it is important to take into consideration the realities of developing new products.

Furthermore, without Mr. O'Conner's efforts in the Professional Baseball Agreements, we might not have the beneficial structure between the MLB and the MiLB that is currently present. His negotiations seemingly saved the relationship, something that simply cannot be done currently with artificial intelligence.

* * *

It is hard to determine the balance between humans and artificial intelligence in sports. There is certainly still a need for human interaction in the current landscape, but as we move forward, can humans keep up their necessary involvement with technology?

"The existential risk is just way overblown. It's much more likely that an asteroid will strike the Earth and annihilate life as we know it than AI will turn evil. It's just improbable and hundreds of years in the future."—Oren Etzioni, CEO of Allen Institute for Artificial Intelligence

People use it more than they realize. In sports we have many cases of weak AI, but no cases of what is called artificial general intelligence. What we have is limited target cases for AI. It is quite frankly just as smart as a 5th grader when it comes to these narrow tasks.

Popular culture misconceptions. Hollywood's portrayal of the AI hype doesn't really capture what is going on with AI. The truth is that behind any AI program that works is a huge amount of human ingenuity and time. The amount of time needed to create AI programs eliminates the possibility of programs taking off like in the movie Ex Machina. The fact that AI will take over simply has no foundation.

Rational expectations. People's expectations of AI are over-blown, creating a stigma that AI is extremely dangerous. Some predict that lawyers will be out of work in a decade. So Etzioni and the Paul Allen institute put on a competition allowing for contestants to submit AI technology that completed an 8th grade science test. Not one team was able to achieve higher than a 60% on the test.

State of the art AI cannot do better than a D on an 8th grade science test because it lacks the ability to understand complex questions which involve mixing scientific knowledge with common sense. But what is interesting is that machines can beat the world champions in almost every board game that exists.

The difference lies in the domain of the information. Even in a game called Go, that has $10\backslash^{170}$ moves, the machine can beat the world champion. Even though the domain is gigantic, it is still finite. Language involves so many variables that is nearly impossible to master.

"*Humans have no chance in these discrete, ultimately artificial board games. The machines are going to win.*"—Oren Etzioni

* * *

If a machine struggles to understand the infinite domain of language, how can we expect a machine to replace athletes altogether? Not only is the software required to do so, but also the hardware to replicate the infinite domain of human interaction and emotion.

From Mr. O'Conner's negotiations to Mr. Stetka's relationships with Orioles alumni, it is clear to see that certain elements are integral to the continuation of sports. Replacing these human interactions in sports will likely not be phased out in the near future. Of course it is a possibility with artificial intelligence, but I do not see it improving as drastically as to handle negotiations between leagues.

Another thing that I doubt will progress heavily in the coming years is a machine's understanding of language. IBM's Watson is an advanced case of a machine's understanding of language, but most machines are still in a phase of learning.

This goes for the physical element well. As we saw with AlphaGo, there certainly is a precedent for machine's dominance in games. But playing a board game is significantly different than physically playing a sport such as soccer. Quite simply the hardware behind this technology has not been developed to a point where it can compete with human athletes.

While AI can be a great help and advantage going forward, there are certain areas where the human element is utterly necessary for the time being.

When discussing the possibilities of artificial intelligence, it is a slippery slope to say what is and isn't possible. There is little precedent for an impact from artificial intelligence on the physical play of a sport, yet with the rapid developments of AI just in the past 20 years, it is not impossible. I will discuss more on this in the final chapter.

CHAPTER 3

PREDICTIVE STATISTICAL ANALYSIS

___

"You get on base, we win. You don't, we lose."—Billy Beane

The rise in the popularity of sports analytics can be traced back to the book *Moneyball: The Art of Winning an Unfair Game* by Michael Lewis. Also made into a movie starring Brad Pitt and Jonah Hill, the book details the rise of the Oakland Athletics out of mediocrity, led by their General Manager Billy Beane. Beane was able to strike a balance between scouts' opinions, character, and statistics in order to lead his team to success.

His focus on obscure performance statistics to make personnel decisions ultimately led his team to win 20 games in a row, the longest win streak in American League history at the time.

Beane created a statistical revolution that has continued in all majors sports to this day. Relying less on gut feel and simple statistics, teams are able to analyze talent in a more efficient and accurate way.

But the revolution doesn't stop in the MLB. Moneyball 2.0 is happening beneath our feet as we speak.

Take for example the Philadelphia 76ers, and what has been dubbed "The Process."

In May 2013, the Sixers hired Sam Hinkie to be the General Manager of their organization. Hired for his analytics acumen, Hinkie began to execute his vision for his team. Even after a 10-72 season, the Philly faithful remained steadfast in their belief in Hinkie's vision. Because Hinkie was able to offer a concrete plan based on quantitative analytics, the Sixers stuck with him.

Before Hinkie took over, the Sixers were a mediocre team, full of mediocre players, destined for mediocrity. But Hinkie set in motion one of the greatest turnarounds the sports world has ever seen. By tanking for a few years, Hinkie was able to bring in players that he projected to be superstars. Tanking is a process in which a team endures losing seasons in order to secure high draft picks. Great players like Ben Simmons and Joel Embiid have proven that Hinkie was right all along.

Ben Simmons is projected to be rookie of the year with the numbers he is putting up in the 2017-2018 season.

In a sport where draft picks can easily be busts, Hinkie managed to successfully draft outstanding players using his statistical background.

The Process is *Moneyball* all over again. And it will continue to grow in popularity.

* * *

Another human has lost to an IBM supercomputer powered by deep learning. World champion chess player Garry Kasparov lost to IBM's supercomputer in 1977. You may remember that Ken Jennings lost to IBM's Watson in Jeopardy in 2010. Recently, in a game called Go, the best player in the world, Ke Jie, was defeated by a computer system called AlphaGo.

The total number of moves in Go is approximately $10\backslash\verb|^|170$, compared to $10\backslash\verb|^|80$ atoms in the entire observable universe. This defeat is particularly tough for humans because some theorized that Go was too complex for a computer to master.

The Economist explained concerning a game against Lee Sodol in March 2016 . . .

"Until Mr. Lee's defeat, Go's complexity had made it resistant to the march of machinery. AlphaGo's victory was an eye-catching demonstration of the power of a type of AI called machine learning, which aims to get computers to teach complicated tasks to themselves."

Like Elon Musk's robots, AlphaGo was taught the tactics and strategy originally by its programmers. Improving its play with every game, it began playing millions of games against itself. Its programming detailed a reward function which essentially let AlphaGo know what its goal is. The trick is to figure out how to reach the end goal. By having intermediate goals, AlphaGo learns what moves are good and bad.

Ironically, following Ke Jie's defeat, he went on to defeat 22 straight opponents after studying the moves of AlphaGo.

This is not the first time that big data analytics has led to insights that could be applied to human activities previously thought to be unlearnable. Defensive shifts in baseball are a great example.

The Pittsburgh Pirates were one of the first teams to implement shifts into their defense, a move that is now commonplace in modern day baseball. Their analysis of opposing players' hitting patterns found that many hitters hit to one side of the field, the pull side. This strategy led to fewer runs scored for opposing teams and more ground ball outs.

This strategy is completely counterintuitive to human thinking.

We think: *"What are we going to do with all the open parts of the field?"*

The important thing that the Pirates realized that it is very difficult for hitters to change their natural swing. Their strategy was successful and was adopted by every other team in baseball. Now we will see every team have a different shift for different players, sometimes having the second baseman in short right field.

The difference between the Pirates and AlphaGo is time. The Pirates spent days and weeks analyzing data in order to come up with the idea of shifting. It took AlphaGo nearly two days to determine the equivalent of field shifting for Go.

The fact that the computer came up with this disruptive technique in just two days proves its computing power. And beyond its computing power, the computer can generate its own disruptive techniques for reaching an end goal because it doesn't think like a human.

Humans should leverage the power of machines to come up with disruptive ideas to gain competitive advantages in sports.

* * *

Sports lend themselves to quantitative analysis.

As I mention in the third chapter regarding computer vision, sports analytics requires large amounts of data.

The British Machine Vision Association and Society for Pattern Recognition explains computer vision succinctly . . .

"Humans use their eyes and their brains to see and visually sense the world around them. Computer vision is the science that aims to give a similar, if not better, capability to a machine or computer."

Sports data is generally derived by computer vision technology, but this chapter covers the specifics in which we are able to make sense of the overflow of data.

The use of data in professional sports has grown tremendously in the past decade. The large amounts of data available to us make it increasingly more difficult to capture all of the relevant data points available. Analyzing and creating useful takeaways from this data is even more complex than selecting useful data points. Big data is a huge section of Artificial Intelligence, especially in sports where statistics play a huge role in the success or failure of a team. Providing coaches, scouts, and players the relevant data to maximize performance is a challenge with the advances of technology and the ways in which we gather data.

While player tracking technologies have been around for a while, the context of this data is becoming increasingly important as the days march on. Teams are starting to use the power of AI and machine learning to amplify their own human potential of gathering and analyzing data. In one such game, a team can gather millions of data points to be analyzed, which is clearly beyond the capabilities of human beings.

AI not only involves collecting and analyzing data from real life events, it also incorporates simulating and analyzing data from large quantities of events in order to create more data. And the larger the amounts of data the better. Earlier I talked about the need for more and more data in order to create slight advantages over a team's competition. The added influence of simulated data can help with selecting players, aiding coaches in their decisions, and even trainers to help with recovery after a game. It can affect almost every aspect of any sports team's strategy and the way they approach a game or even a season. The possibilities currently seem endless with the large amount of data we've had from over the years.

The advantage of machine learning hinges on the ability to make faster and more accurate decisions in the team's daily activities. It can have an effect on nearly every facet of a team's operations and therefore lead to a significant advantage over opposing teams.

With such advances in AI and machine learning in the recent past, the technology is only going to improve and therefore improve the overall quality of the professional sports we enjoy. In sports, where incremental gains are inherent to the success of a team, machine learning and big data are vital in the improvement of professional sports as we know it.

As our technologies become more sophisticated off the field, they have great potential to improve performance on the field.

* * *

Ever since this incredible run by the Athletics, people have been enthralled by statistics in sports and some people had enough foresight to do something about it.

"I'm going to do everything in my power to help this club succeed."—Dan Duquette

The change in evaluation of players by general managers like Dan Duquette resembles an AI-based change that exists in baseball. If Duquette and other GMs across professional sports want to succeed, then they must investigate the new wave of

AI-based predictive analytics.

I wanted to research and learn more about its impact on the sports world going forward. Sports have changed so much over the past few decades and I know it will continue to change going forward.

Dr. Patrick Lucey is the Director of Data Science of STATS, a company whose goal is to revolutionize sports analytics with its AI technology. As of 35 years ago, STATS was just collecting sports data and publishing its findings. But seeing an opportunity to change the sports analytics game, it pivoted its business plan into other areas, aided by new computing innovations such as improvements in computer vision.

Now STATS has upwards of 800 clients that includes Google, Microsoft, ESPN, Snapchat, FIFA, PSG, Chelsea Football Club . . . just to name a few.

You might be asking yourself, "Woah that's an impressive list. But what does STATS actually do?" I'm glad you asked.

Dr. Lucey draws a distinction between structured data and unstructured data. Structured data deals with real numbers and while difficult, is easier to analyze than unstructured data. A large part of his work helps answer tough, more complex inquiries like tracking the various movements of players or balls. The ability to analyze things such as body pose require extremely granular data.

Handling granular data can answer questions like "Under what circumstances is a ball handler likely to take a shot or pass to a teammate?" STATS ability to handle problems such as these has provided them with much success. But like any good leader, Dr. Lucey is looking to better his product.

The transition from analyzing granular data to predicting very specific player movements and capabilities is one that Dr. Lucey and his team are taking on. The field of tracking data is close to being maxed out because "in terms of tracking data, we're never going to have enough." While they are obviously able to provide clients with a winning edge given their track record, they are still looking to expand their data sets. Only through expanding their data sets can they begin to do better predictions based on new examples generated from new data

With the large amounts of data obtained by STATS, you might not expect this to be the case but it is. The need for more and more granular data to "synthesize and generate new examples" is the next frontier of sports analytics, Lucey says.

With their current method of tracking data, they are employing machine learning to learn movements of players and formations to be able to predict movements. Obtaining more precise predictions of their movement lies in a whole separate field of AI called deep learning and specifically neural networks. Synthesizing this new data requires techniques that can learn non-linear relationships between data points.

Lucey doesn't go into detail about how exactly the neural networks are created but it involves creating decision trees, various forms of clustering and unsupervised learning—topics I will not cover in this book. It seems as though deep learning is on the rise again. Recall that Dr. Jeremy Bolton mentioned the recent rise in popularity for deep learning and neural networks. If we can figure out how to fully utilize neural networks, we may reach a breakthrough in creating more data to further AI based sports analytics.

This technology is ground breaking in and of itself, yet STATS is looking to get even more accurate and specific information to help teams succeed in not only team performance but fan engagement as well. They have realized that helping with fan engagement can be just as profitable for them. AI could provide a solution to falling fan attendance and viewership over the past few years.

* * *

The MIT Sloan Sports Analytics Conference remains a major forum of discussion for the increasing role of analytics in the global sports industry. It has become a place in which industry professionals, leading researchers, and executives gather together in order to foster growth and innovation in this arena. The winner of the research papers competition was Patrick Lucey, a recurring character in this book. He wrote a paper entitled "'Body Shots': Analyzing Shooting Styles in the NBA using Body Pose." Along with Panna Felsen, a colleague of Lucey's, they developed a novel attribute-based representation of a basketball player's body pose during his three point shot. While this piece gets very technical, I will describe their work and its implications in the Artificial Intelligence sector.

The attributes that they analyze revolve around not only the jump and release but also the attributes describing pass quality, direction of movement, and footwork. Every NBA player has a different shooting style, some prettier than others, but having a "pretty" looking jump shot is not an objective measure that can be analyzed. The objective measure that they aim to analyze require trajectories of players' body pose.

Before recent advances in computer vision, players would have to wear a motion-capture suit in a very controlled setting, making analyzing player movement an extremely difficult and unrepeatable task. With recent improved automation detection of a person's 2D body pose with a camera, we have eliminated the need for such a contrived setup. Now with this technology, Lucey and Felsen set out to discover if body movements did indeed correlate with a made or missed shot. The three point shot allows for a consistent analysis within the game context and reduces the variability of body pose during the shot. Therefore, it allows for more accurate data.

They then divided the player's body pose during the shot into 5 categories: before receiving the ball, with the ball before the shot, immediately before the shot, the jump and release of the ball, and after the landing. Then each category was given different attributes explaining the body position of the player at each instance of the shot. For example, while a player has the ball before he releases the shot, the attributes are pass quality, pump fake, and dribble—each having a possible effect on the success of the shot.

Without getting too far into the weeds on their analysis, they also were able to analyze other player's movements in relation to the shooter. These movements help identify the difference between open shots and tough shots which is deemed shot difficulty. This is made possible by training the camera to identify made and missed shots using spatial and temporal features captured by their camera technology.

The spatial features were comprised of player and ball x,y positions and the angle and distance between each player and the ball. Using training data from the Golden State Warriors regular season games, they trained their model to predict the success of a three point shot, given a total of 1500 total attempted three pointers.

Finally, they performed a case study on Steph Curry, one of the most successful three point shooters in recent memory, and compared him to his teammates. Given all the categories, characteristics, and attributes they were able to analyze his body pose. The numbers indicated that a higher percentage of his shots are off a dribble and not a pass, off-balance, and he moves a lot more than most players.

This analysis is extremely relevant today due to the advent of the necessity of a three point shot in the NBA game. Just recently, the three point shot has become an integral part of every NBA team's game, and quite frankly, a necessity for success. The case study of Steph Curry is especially interesting because it delves into the success of one of the best three point shooters in NBA history.

While it is an extremely interesting article, that I suggest you read, it wouldn't be possible without the AI technologies of computer vision and machine learning employed to train the statistical models to analyze body pose. These technologies are truly the cutting edge of sports analytics.

* * *

Basketball lends itself to computer vision enabled statistical analysis. The small size of the court makes the creation of spatial models less difficult than, say, a soccer field. This allows us to focus on more accurate statistical analysis. Beyond just evaluating statistics, we can create predictive models that can attempt to predict when a certain player will pass or shoot the ball. This is not a new concept, as I've mentioned it before. Each probability of shot or pass relies on so many categories of statistics that it is a task requiring a lot of data to train the program to learn such patterns. Analyzing behaviors in team sports is more complex than just taking specific individuals into consideration, because patterns of the group matter. Because of these challenges, most work in sports analytics has not considered all possible effects on the probabilities.

Dr. Patrick Lucey and his team, in their paper entitled "Learning Fine-Grained Spatial Models for Dynamic Sports Play Prediction," set out to make in-game predictions of future events over a large selection of in-game scenarios. Their paper goes way beyond my statistical knowledge, but the overall thesis of the paper includes prediction of passes and shots of multiple NBA stars based on their position on the floor.

The spatial models Lucey and his team examine would go beyond most people's statistical knowledge, but the theme is clear. Computer vision not only allows us to track players, but it also allows us to do much more than track players. It gives us many different data points to consider when we examine this advanced data.

The crux of having ample data points is deciding which ones to combine and examine to build these models. Dr. Lucey is on the cutting edge of predictive analytics using artificial intelligence. Others will soon follow suit with papers and models analyzing different data points, drawing different conclusions.

* * *

As the cost of data gathering and AI analysis goes down, it is easy to imagine the lasting effect it will have on not only the professional sports world, but the amateur sports world as well. Youth sports will greatly benefit from advanced analytics like ones that the 76ers and Sam Hinkie have access to. Ultimately, youth sports could get more competitive and selective, resulting in better players rising higher in their sports careers, and the potential for discovering athletes who would otherwise go unrecognized.

Hudl, a highly popular software company that offers tools to edit and share video and analyze stats for amateur sports, now has a product called Sportscode. It allows one to tailor flexible performance analytics specifically for your team. Using computer vision technology, it is able to gather data and analyze it.

Sportscode is just the start of advanced analytics making its way into amateur sports. I expect that Hudl and other companies will gain success as data gathering and analysis become more cost effective.

CHAPTER 4

COMPUTER VISION

———

"What is it about our sports world, and society in general, that wants to know about something before it happens? I'm OK knowing about it when it happens."—Buck Showalter, manager of the Baltimore Orioles

A lot of people focus on staying in the present. With artificial intelligence we are more concerned with what is about to happen. And the technology and the data is there. We just need to find out how to use it.

If it hasn't become clear, the Baltimore Orioles are my favorite sports team. Their manager, the legendary Buck Showalter, was quoted saying the above words. The way the field of data analysis is trending, Showalter will learn that it will become important and even commonplace to have predictive statistics on opposing teams.

Techopedia says . . .

"Computer vision is a field of computer science that works on enabling computers to see, identify and process images in the same way that human vision does, and then provide appropriate output."

Computer vision is a popular field in which machine and deep learning can be applied. Similar to image processing, computer vision is especially useful in the sports arena when used to track players' movements on the field or court. With the ability to film this intricate data, we can gather much more data, a necessary component for training machine learning algorithms.

Basically when we collect this information, we are able to feed the data into machine learning algorithms and neural networks to do a number of things. A major use of this data can be for predictive analysis of a player. Having specific recorded data of players' actions on a court can help a team assess talent more effectively, or an opposing team play to a player's specific weaknesses.

Computer vision as a form of AI contributes significantly to many of the topics I will cover in this book. It directly influences and creates a majority of the data needed for machine learning algorithms in all facets of the sports industry. It plays a direct role in predictive statistical analysis which is covered in the previous chapter.

* * *

My experience with CV. One challenge I have faced in my life was getting cut from my travel baseball team at the age of 10. After only one full year on the team, I was informed that I would not be offered a spot on the team. For someone whose life revolved around baseball, this was crushing. But this adversity contributed to the growth of my character and diligence.

My mother, who knew how much this meant to me, took me to the batting cages every day over the summer to practice my swing and prepare for tryouts at the end of the summer for a different team. My mom was dedicated to getting me there every day, even if it interfered with her schedule or she was tired. My parents even got me lessons with professional instructors. They introduced me to a primitive video app that was able to break down my swing into every little detail. Through this app, I was able to diagnose the problems in my swing and learn to correct these errors.

Well not surprisingly my hard work in the cages paid off when I was offered a spot on another team. The irony is that when the new team I made broke up four years later, I was offered a spot back on my old team that had cut me. I have faced other adversity in my life but this is the one of which I was most proud.

Although the video app required no artificial intelligence, it did require a camera that was able to break down different features of a swing. I didn't know it at the time, but this was the prequel to computer vision.

* * *

Professor of Computer Science Dr. Jeremy Bolton is a specialist in the field of computer vision, a specific form of Artificial Intelligence. His expertise in the field is what drew me to interview him and he provided me with extensive insight into the usefulness of computer vision.

He first got involved with Artificial Intelligence in an image processing course at the University of Florida. This course actually incorporates both computer vision and machine learning, a popular combination of tactics when attempting to use computer vision effectively. His first project exemplified the usefulness of such a technology. His project was intended to detect landmines in the field of battle, a powerful and potentially life-saving project. Dr. Bolton says this may have been his most rewarding work because landmines can have such a dangerous impact on our soldiers at war. This is especially relevant since in 2015 landmine deaths across the world hit an all-time high with approximately 6,461 victims claimed.

Beyond his computer vision course, Dr. Bolton thoroughly enjoyed the theory of AI as it is an elegant form of computer science.

Furthermore, we delved into the challenges for AI that he has seen over the years. One of the greatest challenges that we face in AI today is modeling human behavior. According to Dr. Bolton, modeling human behavior is challenging and nearly impossible to do. Having done some work with mapping personality models to feature spaces, his experience has only reinforced this conclusion. Given the intricacies of the human mind, accurately representing features like emotion and complex thought prove a difficult challenge in AI today. The ability to characterize human behavior remains one of AI's greatest challenges.

This notion obviously calls into question the future of Artificial Intelligence and its ability to make advancements. There are more complicated approaches to the idea of modeling human behavior that I will not discuss in this book as it may get us sidetracked from its application in sports. The different techniques we have seen implemented in sports are useful, but the challenge of modeling human behavior could indicate that there is a limit that AI might reach in its ability to enhance the sports world. But this does not mean that it is impossible. Dr. Bolton and I continued our discussion of AI and how he has seen it change since his years in college.

For him, the most interesting change in AI has been the rise and fall of neural networks as a valid technique to model the human brain and nervous system. Networks generally have the ability to model the human brain and attempt to "think like a human." For example, our brain is trained to read and write the English language. Over years of studying and training our brain learns what each letter and word means. The idea behind neural networks is similar. Through the repetition of "training examples," a neural network will attempt to devise a set of rules that can be applied to these specific examples. In one case, a neural network can be fed different handwriting examples and will establish a number system from what it can see.

Getting away from the technicality of neural networks, they were previously dismissed by computer scientists after they were proven unable to solve the specific XOR logic problem. Years later it was proven that neural networks could actually solve the XOR logic problem, so they have been on the rise ever since. This rise can be classified as the advent of deep learning, the use of neural networks in modeling human behavior. Dr. Bolton mentioned that controlling these neural networks is tough and could pose problems in the future.

Finally, we got around to the topic of sports. There are many articles on the recognition detection problem in computer vision. He mentioned an article that discusses ways to automatically identify football formations and provide statistics just using a camera linked up with machine learning. This data gives us a higher probability of guessing which play is coming from the defense. He finally mentioned identifying tracking of pick and rolls to predict plays on a basketball court.

This leads me to a Ted Talk which I watched that was given by Rajiv Maheswaran.

* * *

In an age where we are incredibly good at capturing information about ourselves, the detailed information we can gather is incredible. We have apps that can track our sleep cycles! We have other apps that can tell us how many steps we took in a day. The limit to which these technologies can capture information seems endless.

As it turns out, and I've said before, sports offers an amazing opportunity to gather huge amounts of data, whether it is statistics or something more. In a sense, what this comes down to is reducing ourselves into moving dots. In a recent

Ted Talk by Rajiv Maheswaran, he says that with computer vision we have access to huge amounts of raw data based on these "moving dots." These moving dots being basketball players on a court, for example. While a coach cannot see every movement of his players at each second, a machine actually can. This is what gives teams the advantage of using machine learning through computer vision.

Slowly we have been able to train the machines to recognize simple plays. For example, in basketball it can recognize shots and rebounds with ease. The tough part of using this computer vision is adding a human element to the machine. One could easily describe a pick and roll, and even point out a pick and roll on any given play. But a machine has a tough time learning what exact movements are pick and rolls and what movements simply look like a pick and roll.

The only way to train the machine to learn a pick and roll is through more and more data. It all comes down to data. With humans help, the machine can begin to recognize that yes, that is a pick and roll. Or no, that was a slam dunk. But through its machine learning algorithms, the machine can figure out with great accuracy whether a pick and roll occurred on a particular play.

This raises the question can machines know more than a coach. Maheswaran says yes and it makes sense.

Using spatiotemporal features, we can break down almost every aspect of every movement and create statistical models using machine learning. A large amount of statistical models, with or without computer vision, use machine learning to produce results. He says the average NBA players makes a shot 49% of the time. But with our new information, we can break down shots into two categories: the quality of the shot and the quality of the shooter. Each of these categories can be further broken down into categories that help identify each quality. Therefore you can have a good shooter who takes bad shots, and a bad shooter who takes good shots. This helps with not only player development and but also talent evaluation.

It is pretty clear that Lebron James is both a good player and one who takes good shots. But what about a lower caliber player? How do we decide which player to give a max contract to? Data derived by computer vision can enhance the statistics with which teams evaluate talent and ultimately lead teams to be significantly more successful.

Reducing players to moving dots seems silly, but actually can be highly effective.

Maheswaran closes by saying that you do not need to be a professional player to track movement, opening the door for this machine understanding to apply to amateur sports as well. Even outside of sports, we can track every movement in our lives which could provide us with insights we wouldn't get without this technology.

* * *

The Olympics is the oldest form of competition still in existence. The first games were held in Athens, Greece and began in 776 BC. Now, in 2017, the Olympic Games are the most watched sporting event on television. With over 200 countries that send athletes to the games, many countries cheer their countrymen and women on in their respective sports day in and day out over the 2 weeks that they run. This type of attention also draws a lot of controversy, whether it is during a competition or the judging of a competition. Surely over the countless events that occur over the span of the games there are bound to be human errors in the judging of events such as diving and gymnastics, two sports that rely heavily on the need for accurate judging of the competitors' skills.

One instance of this controversy occurred recently in the 2012 Summer Olympics men's gymnastics competition. Japan was promoted to the silver position after they filed an appeal which said their gymnast Kohei Uchimura's final pommel horse performance was judged incorrectly. This bumped Ukraine out of the bronze medal position, devastating their athletes and supporters alike. The correction to the scoring was ultimately accepted by all parties but left disappointment in the hearts of the Ukraine team.

In an effort to avoid mistakes such as these, the International Gymnastics Federation (FIG) is planning to use AI technology to assist with scoring the Tokyo 2020 Olympic Games. Fujitsu, a Japanese technology giant, is developing a 3D sensory system with computer vision that will help make scoring easier, assist coaches and athletes in training, and offer broadcast viewers an in-depth coverage of the event. They are currently focused on tracking the body position on specifically the vault apparatus. It is currently analyzing and collecting data from previous vault runs to advance its accuracy. The goal is to have a product that will assist judges in scoring and speed up scoring delays to ensure every athlete's nuanced technique is appreciated and scored correctly.

Another advantage to this technology is that it would not get fatigued. As an avid fan of the Olympics, I have watched a few gymnastics competitions over the years. Not once have I considered the fatigue of the human judges throughout a day's competition.

"A judge must work for eight hours per day—does that allow the mental capacity to remain coherent? It's not possible to maintain a coherent mind of criteria. Only the computer does," *says former FIG president Bruno Grande.*

Eliminating mistakes and fatigue are huge factors in considering the success of this technology, but it could provide some downsides that I had not considered either.

One such drawback is the new territory that is being entered. For a sport that has had scores calculated based on human judging for centuries, judges are wary of how the technology will affect the scoring system. Little is known about the technology currently, which is interesting considering the grand stage it will be displayed on. Without years and years of testing, it has certain gymnasts and judges questioning the accuracy of the product.

"The judges have a very tough job on their hands and sometimes they can get a lot of stink for what they do," said Great Britain's Olympic champion Max Whitlock, who was surprised by the pace of the technology's introduction. *"It would have to be absolutely perfect. Something like that would be so new to gymnastics it would have to take years and years of testing. It would be very exciting. But it will be a long time in the future, probably not my generation."*

While the technology will be tested at the world championships in Qatar, an almost equally competitive stage, it could prove to be unreliable and in certain cases just wrong.

Another drawback is that this technology would discourage

innovation. Gymnastics is a unique sport that is known for its rigid technique and the ability to show flair in a routine. Gymnasts are constantly pursuing new skills to differentiate themselves from their competitors.

Olympic great and the first woman to record a perfect 10 on a routine in gymnastics, Nadia Comaneci, reverberates these concerns: *"Gymnasts are known for pushing the skills, looking for new angles, turns, points— so what happens when someone comes along with a totally different routine that has not been seen or registered by the computer. How would that be judged?"*

In an attempt to eliminate inherent bias in judging, we could eliminate the creativity of athletes that makes the sport so outstanding.

Another such concern is the lack of security for such a product. Hackers theoretically could manipulate the algorithms responsible for judging the games that mean so much to so many people. Athletes have worked all their lives to get to the Olympics and could have their hard work eliminated by a manipulation of the technology.

This drive to incorporate such advanced technology is part of Japan's effort to make the 2020 games the most revolutionary and technologically advanced Olympics game ever. Could Japan be too aggressively pursuing this goal and ultimately sacrifice the integrity of the competition just to eliminate human bias? It remains to be seen.

The possibility of eliminating the need for judges is possible but unlikely. Having a score calculated by just a camera and a computer is an unlikely scenario given the development of the product thus far. Comaneci points to an extremely relevant drawback of introducing this technology to the sport. It could end up restricting the evolution of gymnastics as a sport.

She leaves us with a relevant quote: *"I don't think it will be possible to ever replace judges. Gymnastics is too complex, there are so many skills and nuances in every routine. But I like the idea of the technology. It's OK to try, why not? This is what is happening in all sports now, technology is changing our experience with it."*

* * *

Another example of the introduction of AI into sports comes from MLB Advanced Media. They have risen out of the changing landscape of MLB baseball, beginning with the Oakland A's rise out of mediocrity in In a game that has evolved so much in terms of the way baseball operations are handled, statistical analysis drives this engine of change. Just recently, MLB has introduced fans to their new product Statcast. Even my first impression was a feeling of awe, but Statcast can go way beyond what the average fan might think.

Statcast is an addition to the broadcast of a game or highlights of a game that helps provide fans with interesting statistics that they would not have access to otherwise. Using computer vision, it is able to track pitchers, pitches, batters, and even fielders constantly through a roughly 3-hour game. It can give stats when a player like Giancarlo Stanton, a player for the New York Yankees, hits a home run, a real cracker over the left field fence. In 2017, he recorded the hardest batted ball, clocking in at 120 mph off the bat. It gives stats such as exit velocity, projected distance of the home run, the launch angle and even the angle at which the ball left the bat. It can also track players' first step, measuring their reaction time to a batted ball. It tracks Adam Jones, the center-fielder for the Baltimore Orioles, and his top speed to track down a ball in the gap. It is really an incredible tool that gives fans an extra look into the raw power and speed these players exemplify. Essentially it gives fans quantifiable measures for what was previously not quantifiable. But even though Statcast is computer vision, is it really AI? The answer is quite clearly yes.

Imagine the amounts of data that we have access to over a course of a 162 game season, in addition to Spring Training and playoffs. Claudio Silva, a Professor of Computer Science at NYU and co-developer of Statcast, says there are approximately 700,000 at bats during any MLB season. Just one game can generate up to 80 GBs of compressed data, 7 terabytes uncompressed. For perspective, the laptop I am writing this on holds about 16 GB of data. The amount of data collected from just one given season is massive and enough to start applying machine learning techniques to it. By tracking every movement of every player and the ball throughout the game, we are able to apply machine learning statistical models to it.

It has already changed how coaches evaluate and train their players and how fans watch the game. But Silva believes it can do much more. He is employing deep learning to reveal minute details of player behavior and game patterns. It can even give Silva and his team the ability to make predictions about some aspects of the game. Through an established deep learning network for a player's needs and tendencies, coaches could use this product to detect when players are fatigued or even injured. With arm injuries playing a huge factor in pitchers these days, this could revolutionize the way teams tend and care for our pitchers.

On top of the Oakland A's sabermetrics, Statcast employs these big data and machine learning tactics on the nearly 1.5 million plays of data collected over just 2 seasons of baseball. The Statcast database is so refined that each play has textual descriptions, video clips, outcomes, and player positioning movement data.

The improvements that could be made to this current system, Silva says, are costly but would be effective. Buying more cameras in order to analyze more detailed player movement is the next step, but currently the computing capabilities are not quite up to speed. For now they will continue to develop and refine their deep learning networks for each player and attempt to derive more meaningful statistics from these minute data points.

Baseball is a game that has been played for nearly 2 centuries so it is interesting to see how it can still be greatly improved by not only technology but also specifically big data, machine learning, computer vision, and deep learning. Baseball can only go up from here.

Silva says fittingly, *"This is a game that's been played a long time; what surprises me is that we're still able to make it better by using technology."*

The improvements that could be made are clear, but they face some financial and technological limitations. The more technology Statcast and Major League Baseball Advanced Media can obtain, the more data they can provide for organizations and fans. Having seen the revolution that Moneyball brought about, the similarities to the rise AI-based statistical analysis are remarkable.

CHAPTER 5

SPORTS BETTING

———

The Final Day. May 5, 1996.

AGUERROOOOOOOOOOOOOOO

Down 2-1 to Queens Park Rangers in stoppage time, Manchester City needed two goals to win the Premier League title, the top soccer division in England and one of the best in the world. After equalizing in the 92nd minute, Sergio Aguero slots one into the back of the net to put Manchester City up 3-2, stealing the title away from Manchester United, their cross-town rival, who only seconds before had tied Newcastle United. A win would've secured a title for Manchester United.

Even though I cannot claim myself to be a huge soccer fan, I still get the chills every time I watch the YouTube video.

The announcer screams AGUEROOOOOO as Aguero runs around the field in celebration, ripping his shirt off in the process.

But what if we had a way of accurately predicting the winner of the title, even before it happened? Well, it has become a possibility with Artificial Intelligence. AI is making its way into the sports gambling industry, and certain companies utilize AI to correctly predict winners.

* * *

The sports online global gambling market is a massive industry that brings in astonishing amounts of revenue. Recently, it was valued at about $39.8 billion dollars. Yes, it is hard to estimate because regulations differ around the world. But, there is no denying its sheer power in the global market. Somehow finding a way to "game" the gambling market could be worth billions of dollars in the future. A firm named Stratagem is attempting to utilize neural networks in order to predict the outcomes of soccer games.

A lot of my research has consisted of applications of machine learning based on large amounts of raw data. Well, data seems to be a common thread here, but neural networks are not utilized as often as machine learning.

Stratagem, an AI company attempting to use their technology to predict outcomes of soccer games, is testing the waters to see if this is viable. They are using large amounts of data to train their machines to track things like scoring changes in soccer games.

"Think about oil in the ground, all of this in various locations. It's the same thing as data."—Andreas Koukorinis, the founder and head of trading at Stratagem

Previously involved in quantitative trading at a hedge fund, Koukorinis left his $3.3 billion hedge fund to apply his trading knowledge to sports betting. His commitment to this idea is impressive.

"2013 was me in my living room with my wife being like what are you doing with your life?" he recalls. "You used to work in finance, we used to fly first class, and now you're sitting here in a t-shirt in our living room with two guys, you guys are barely speaking—this is crazy. I said, no, no, there's something here."

Their process involves gathering large amounts of data, then "crunching it." Sounds simple, right? Strategem's program can read all sorts of different data outlets, such as Twitter feeds, crowdsource videos, and market data. It then is able to assign a higher weight to more trusted and reputable sources.

With these sources, the end goal is to identify "alpha" in the market. Alpha is a performance measure that gauges investment performance based on a market index. For those of you who work in the financial market or have friends who do, alpha is mispriced odds where Stratagem has a better chance of winning a bet. The program is even able to place bets, both before and during games.

"For us, it's really about having access to data that comes from multiple sources and of different textures and having the backbone of the overlay to be able to analyze them . . . That's really the edge."—Koukorinis

Stratagem is also using AI to figure out when the best time to place a bet is based on the given odds at the time. This involves predicting where the line, the odds one team will beat another, will move after it is released. Similarly, they attempt to predict the starting lineup which can be a good indicator of where the line will move.

They also are able to transform the pitch into a 2D map of the soccer game in order to analyze things like formation, positioning, and spacing. The whole model revolves around the fact that goal scoring opportunities are a good statistic to predict outcomes of games. This raises the question, are there other metrics that facilitate the prediction of games?

Koukorinis believes this is a viable option to reinvent the sports gambling industry. Their systems are currently analyzing hours of soccer video to create deeper neural networks that make their predictive algorithms more precise.

There is no guarantee that this strategy will work. In 2012, a sports betting company lost $2.5 million after promising investors returns of 15% to 20%. It will be interesting to follow Stratagem in the future to see if their strategy pays off. For now, I remain skeptical. It is proven that trusting your money with the world's top performing stocks on the S&P can be safer than investing in an AI hedge fund, but if this works, the possibilities are endless.

The rise in sports gambling resembles the rise of the equity markets in the 1970s and 1980s. It wasn't a market that people actively invested in and the technology didn't lend itself to having great price discovery, meaning it was expensive to trade. These things were obviously adjusted, and Stratagem is betting on the fact that the sports gambling industry is moving in the same direction.

This calls into question how long the sports gambling industry will last if this technology is successful. Personally, I think that the randomized outcomes of sports are what make it so interesting and exciting to watch. This randomness is working against the mission of Stratagem, but I will reserve judgement until I see their success or failure.

The CEO of Stratagem, Charles McGarraugh, optimistically doesn't believe they are crazy.

"Maybe we're on the line between genius and madness, but I'm pretty sure we're on the right side. It's a big global market. It can be better. We want to be part of this."

* * *

To get a little more technical on how exactly sports betting is predicted using neural networks, I am going to present a project aimed at predicting NFL football game results.

The NFL is a multi-billion dollar business, making this study even more relevant. There are several websites that claim to be able to predict the outcome of NFL games. But which ones can we trust? Which ones are actually correct?

The goal of the project was to create a completely objective, statistics based system for predicting the outcome of NFL games. The problem with creating statistical models to predict sports game is the intangible aspects of the game.

Why do we choose a neural network?

Teams can win in a variety of ways and can by no means be mapped in a linear fashion to determine outcomes. The problem boils down to a pattern classification problem, which neural networks are commonly used to solve.

There are many similar studies and projects done on predicting sport outcomes, but a common connection between them all is the neural network. Few sports, if any, can be mapped linearly to predict winners and losers.

The first step in creating the neural network was collecting the mass amounts of data needed to train the network. A large data set is required to represent the many ways a team can win. The preliminary test run revealed 5 statistics that were the most predictive. They include total yardage differential, rushing yardage differential, time of possession differential, turnover differential, and home or away.

From these statistics, two prediction sets were created: one based on team season averages and the other based on average of the prior 3 weeks. The latter giving a better picture of how the team was doing recently, adjusting for teams going on a hot streak.

After training data was fed into the network using methods too technical for discussion, predictions were then output. On average for Week 14 and 15, the neural networks guessed winners with 75% correctness compared to ESPN's average of 72%.

Can we make improvements?

The prediction rate could be improved by adding the human element. If somehow we were able to incorporate intangible considerations, Las Vegas betting lines, and subjective team rankings into our networks, then we may be able to get a more accurate prediction rate.

Another improvement that could be made is including the previous season's data, the more data the better. The ways in which teams win presumably does not change over time if most things remain the same.

A more technical view of sports prediction using neural networks can provide us with insight on how difficult it is to predict a sports game. Stratagem has a long road ahead of them if they are going to succeed. The differing sports could play a factor in their success as soccer may lead to more predictable matches.

* * *

Another attempt to parlay AI into sports betting involves horse racing. For someone who does not normally follow horse racing, the Kentucky Derby is a race that I would not miss for the world. Year in, year out, I still believe that I will bet on a horse that does not win, much less even finish in the top four. I'd imagine this is how it goes for the average horse racing fan, or I'd like to hope so. But even to a horse racing expert or analyst it is very tough to predict the top four finishers in exact order, also called the superfecta.

In 2017, the Kentucky Derby made history. Organizers from Churchill Downs, the company that owns the Churchill Downs track in Kentucky where the race is held, partnered with an AI company in order to handicap the race, or set the odds for betting. Unanimous A.I., the AI startup partner of Churchill Downs, created a technology called Artificial Swarm Intelligence that combines handicapping experts and analysts to make expert picks for the Derby.

Apparently in 2016, the year before the partnership, Unanimous A.I. accurately predicted the superfecta of the Kentucky Derby, a nearly statistically impossible feat. A small bet of $20 dollars would have netted \$11,000 if the superfecta was correctly predicted. This is a task that is hard to do in a small race, and especially difficult on such a huge stage as the Kentucky Derby with 20 horses in the running. A combination of proven experts were gathered to make the "swarm" to predict the top finishers.

Just as a flock of birds operates in a swarm to reach collective decisions to survive, so does this predictive strategy. With the combination of multiple experts picks, Unanimous A.I.'s software uses machine learning to collect the top picked winners of experts to predict the unpredictable. Their software goes way beyond the mental capacities of a single individual to amplify the intelligence of the group of experts.

The founder of Unanimous, Dr. Louis Rosenberg, created the algorithm that allows these experts to benefit from a shared mind.

"While predicting sports always involve a large element of chance, Unanimous A.I. taps the intelligence of groups and evokes the best possible prediction based on the available information. We have seen this work in a wide range of fields, from forecasting movie box-office [projections] to predicting the price of bitcoin. We are excited to see how these handicappers do against one of the most unpredictable of events—the race commonly known as the most exciting two minutes in all of sports."

The software will partner with Churchill Downs and their betting website TwinSpires to allow players to make wagers. In order to eliminate the possible total winnings of thousands of bettors using this software to cash in large bets, TwinSpires will be running a $10,000 players pool that will make wagers based on the results of the Unanimous A.I. Swarm. Bettors can buy a share of this \$10,000 pool with shares of as little as \$10. If the swarm correctly predicts the superfecta, the bettors will win their percentage of the profits. Those who have great faith in this technology will predictably buy large shares of the pool in order to benefit from the swarm technology.

This swarm technology has been used to forecast other events, such as box office projections. It has proven to be reliable and will test the waters in the most watched horse race each year.

This structure of betting could change the landscape of horse racing betting forever. If it can be proven to work in the most unpredictable of races, it surely can be applied to smaller races.

One interesting aspect of this swarm technology is that is still does rely heavily on human influence, and in fact only functions with horse racing experts and their predictions for the race. As the Unanimous technology gets more trained, we could see in the future the decrease in the need for horse racing experts' opinions and possibly the eventual phase out of the necessity for this mob mentality.

It is clear that this is a long way away as this swarm technology is such an inexact science. In 2017, The Swarm projected Classic Example to win the Derby. Classic Example actually finished fourth, and they were not able to predict the superfecta two years in a row. But there is a clear precedent that this technology works and can be improved. In such an unpredictable race, it is impossible to predict exactly what each horse will do because there are so many factors during the race that experts simply cannot account for.

The swarm technology could be applied to a more predictable sport and could get more predictable results. This remains to be seen but is something to keep an eye out for as the years go on.

CHAPTER 6

DATA COLLECTION USING WEARABLES

Andre Igoudala, an NBA player for the Golden State Warriors, and his love for sports goes way beyond his prolific career on the basketball court. Since 2015, the Warriors have been big proponents of wearable technologies to better manage their athletes and keep tabs on their progress. But Igoudala is no stranger to player-tracking technology.

Andre Igoudala's talents extend beyond the basketball court though. He is a golf fanatic. The 2015 NBA Finals MVP has an affinity for golf and the products that go with it. He has used wearables and apps such as Game Golf and Zepp Golf to improve his swing on the course. If you look up his swing, it wouldn't be a stretch to say he looks like a professional.

Game Golf is a product with automatic shot tracking and real-time stats that helps golfers to gain insights into their game and into where they need to make improvements. The product is a sensor on the end of the golf club that sends your swing stats to an app on your phone. These advanced statistics allow players to analyze their swing in ways that they would not be able to otherwise. Zepp Golf is a similar product.

Igoudala believes that wearable products are currently barely scratching the surface of what's to come in the wearable industry. Igoudala consistently makes an effort to keep up with the wearable market and looks for opportunities to invest in products that can scale to a large market. He looks for the ability to combine new technologies and player needs to create a better sports product all around. He hopes his multiple investments will pay off in the years to come.

Wearable technologies are only beginning to gain traction, yet they have proven to be successful for not only the Warriors, but for other professional organizations as well.

* * *

As technology advances, the speed at which we can access data is extraordinary. A quick Google search of "cute cats" returns in under half a second, and you are immediately laughing at a kitten smacking a ball of yarn around. A frivolous search, while fun, is not the kind of data that is important to have in our hands at all times. The use of map applications has revolutionized the way we get around every day of the week. While I am not old enough to remember the days of using actual maps for navigation, I clearly remember the days of printing out MapQuest directions to get to one place or the other. If you missed a turn or hit traffic, good luck!

GPS tracking has become an inherent part of our culture whether we realize it or not. Snapchat maps is a good example of the extent of our GPS tracking use. What used to be an app to send silly pictures to your friends that would disappear after 10 seconds, has rapidly advanced to add features such as Snap maps. This function allows you to know where all your friends are at all times. Creepy, right?

Well, the application of GPS tracking in the sports industry is not only less creepy but highly effective in bringing real time data to elite athletes who would otherwise not have access to such data. These "wearable" technologies implement GPS tracking as their major source of collecting data.

Barbara Kendall, a former sailboarder turned technologist, talks about the advantages not only of GPS technologies but of other AI technologies being currently used in Olympic competition. A 3-time medalist in her 25 year career, Kendall saw this transition like few others have.

Kendall says, *"Artificial intelligence is disrupting the very way we view sports and athletes play the game—and automation is playing a huge role in this."*

The crux of her interview revolves around real-time data from robotic sensors placed on athletes, which involve mainly GPS technologies. For example, sailors have AI technologies implanted in their boats that provide real-time and accurate recordings for race results, speeds, and tactical decision-making. Previously, competitors had to rely solely on their eyes, instinct, and experience to navigate the waters. AI can revolutionize the way sailors receive data on the water to enhance their racing capabilities.

AI technology can not only improve competitors' abilities, but also the accuracy with which judges officiate competitions. This is currently happening in fencing. Sensors placed in fencers gear allow judges to see the exact moment and location of a hit, eliminating inevitable human error or bias. The elimination of error and bias can greatly help to reduce controversy in such a widely viewed stage as the Olympics.

Similar technologies affect competitors and judges, and can greatly enhance the way viewers digest the sports they love to watch. Kendall says . . .

"How fans consume and watch sporting events is constantly

being disrupted by emerging technologies (think of how many more are using mobile devices), and AI is just the latest. Fans can see real time statistics for their favorite players and their competitors, making viewing a more interactive experience."

In the fencing example, viewers can see on their screen at home exactly what the judges see, making them judges themselves while sitting on the couch. In tennis, we can see the speed of an ace on the big screen immediately after the serve. In rowing, viewers can see statistics such as stroke rate per minute immediately displayed on their broadcast. Immediate access to these statistics is becoming commonplace in sports today.

The possible future applications of these technologies are endless and inevitable. Will it be employed to more accurately time races and events? Questions such as these will continue to be answered as technologies become more commonplace in Olympic competition.

This technology is being employed by powerful nations who can afford to dedicate resources to AI. Kendall predicts that as the years go on, AI will also help developing countries who lack the resources of developed countries, to even the playing field at the Olympic level, enhancing the overall quality of each sport.

* * *

The wearable market is just exploding with companies selling their products to customers. Most are able to provide fitness metrics to give feedback for the athlete.

A new sports technology startup, Boltt, is attempting to change the way that athletes use wearables. It includes an AI-enabled personal coach that crunches data and gives personalized instructions to the athlete. In partnership with Garmin, a GPS service, they are working to come up with a customized and informed device that classifies an athlete's training exercise by time, type, and intensity.

It contains sensors that track both cardiovascular and mechanical data. With this data, the Boltt coach, nicknamed B, transforms the data into usable guidance. Using the athlete's goals, B will guide users towards their goal-oriented activities instead of just giving them a load of information that would have to be deciphered.

Other features include guidance on personal training routines, minimization of injuries, intake of nutrition, and recovery time through sleeping habits. Although not currently launched, the Indian startup is looking to gain investors.

The addition of the artificial intelligent coach serves to separate

Boltt from its competition. Speaking of competition, Boltt is not the only company to employ complex AI technologies in fitness wearables.

Vi, an enabled wearable, was launched in June 2016, raising $200,000 in 24 hours on Kickstarter. Vi gathers data by monitoring an athlete's physiology and environment, activity history, as well as reading data from other fitness apps. Its integrated AI then analyzes and adapts the data to provide the athletes with suitable insights through voice or text.

It is currently designed for runners, but plans to be used for swimming and biking workouts in the near future.

Although these certainly have little effect on the professional sports industry currently, they do have the technology to be applied to professional athletes. Just as Catapult, a wearable tech company discussed in the next chapter, transitioned from the professional market to the amateur market, I'm confident that wearables such as Boltt and Vi are capable of making the transition to the professional market. Although, it seems as if Catapult owns a majority of the professional sports market, boasting clients in the Premier League and La Liga.

But Tom Taylor in a recent article in *Sports Illustrated* suggest that the future of wearable technology is not on the wrist, but elsewhere, hidden away.

"The greatest technological advancement to come in the next couple of years could just be invisibility."

Wearables such as Fitbits and Apple Watches are commonplace at sporting events around the world. Leagues like the MLB and the NBA have been forced to approve certain wearable technologies.

The MLB just approved a compression sleeve that tracks the workload of a pitcher's arm made by Motus Global. They also approved a heart-monitoring strap made by Zephyr.

The NFL expanded the use of the Zebra RFID player tracking system.

Wearables have also helped American athletes in the Olympics compete for their country. Mounir Zok, director of technology and innovation for the U.S. Olympic Committee, credits wearable tech as a key part of Team USA's success at the Rio Olympics.

Now an athlete's performance can be tracked in their natural environment rather than in a standardized lab. This reminds me of seeing videos of athletes running on a treadmill with tubes and wires attached to all parts of their body.

The answer to the wrist-worn wearable problem is unclear. The problem results from inaccurate measuring of cardiovascular functions as it is not across the chest near the heart. Taylor suggests that the next step might be intelligent clothing. If devices can be sewn into fabric or hidden inside shoes, they can be placed exactly where they need to be.

"With the smart fabric evolution we will start exploiting the real estate of the athlete's body," Zok says. *"From 2017 onwards we will start integrating the technologies and innovations in a more seamless fashion onto their body."*

Taylor finishes his article with a daunting proposition. He suggests the improvement of computer vision is happening so rapidly, computer vision technology may eventually be able to pick up the metrics that the wearable tracks, rendering wearables utterly useless.

This is an interesting topic to watch as the AI technologies get more and more advanced.

* * *

Wearables in-game. As wearables become popularized in professional sports, the allowance of these wearables in game will become an issue. In 2017, MLB announced that certain wearables were permitted in-game and in practice and workouts. This big step forward for wearables could prove to open the wearable market further.

But issues of player privacy and unfair advantage are slowing their integration into professional sports. With the current wearable technology allowed in game, I would imagine that it will not be a major issue going forward. Companies like Catapult have been approved in many professional leagues around the world to be worn in-game.

The NBA has recently released their smart jersey. With the NBA mobile app, you can scan the chip on the tag of the jersey to see stats, highlights and much more. If this is allowed, there is certainly precedent for the allowance of player performance tracking wearables that teams use to measure performance.

I believe the wearable market is just beginning to grow. As professional leagues see the growth and usefulness of wearables, the wearable market will continue to explode.

CHAPTER 7

TRAINING, RECOVERY, AND INJURY PREVENTION

——

Tom Herman, the head coach of the University of Texas football team, is a football guy. A football guy is someone who lives and breathes football. Someone who can't sleep the night before a big game.

He cares so much about his team and players that he has been known to post signs entitled "Longhorn Football Hydration Chart." What exactly does this entail?

Picture a chart full of different shades of yellow. Yes, Herman expects his players to measure the shade of their urine with the chart. If yours is white to light yellow, you are at "Championship Hydration Levels."

But go beyond light yellow? You could range anywhere from "a selfish teammate" to "a bad guy."

In my opinion, this is taking player monitoring to an extreme, almost intrusive level. There is no denying that there is some comedy to it. But, player monitoring can prove to be useful.

It is a good indication of how far coaches are willing to go to track player performance off the field.

* * *

AI can be used for analyzing statistics to improve player performance—outside of simple talent evaluation. The next step for AI could be in analyzing off the field metrics to assist in an assortment of areas.

Can we use AI to help an athlete recover from an injury faster? Can we get better training results from athletes?

These are all possibilities. The importance of an athletes training regimen cannot be denied.

Chip Kelly, the former coach of the Philadelphia Eagles, is

well-known for his scientific training regimens. While he made an attempt to hide his regimen from the public eye, we can gain a glimpse into how he successfully trained his team not only to perform well in games, but also to prevent injuries.

Eagles' players wore Catapult Sports monitors that measure statistics such as agility and acceleration. These statistics are then analyzed to determine exactly the point when performance starts to decrease in each specific player, creating fewer overexertion injuries. The monitor generates a post workout report that indicates when players will be ready to handle more training.

Kelly invested over $1 million in technology like accelerometers and heart monitors. While this seems like a lot, it is just spare change compared to the \$133 million that the Eagles pay their players in salary.

But, his trust in technology seemed to work.

In his first year as head coach, Kelly led his team to a 10-8 season, which saw them winning their division, the NFC East. Lack of injuries played a major role in their success. According to an injury statistic, they were the second least-injured team in all of the NFL.

Kelly admits that there is no way to prevent a freak accident like a player breaking a leg. But their aim is to prevent soft tissue injuries, such as a pulled hamstring.

"We do that because we think there's a benefit to it. Obviously, the big issues you look at are the soft tissue injures, because those are preventable."—Chip Kelly

And their efforts are not only aimed at preventing injuries, but also improving player performance and limiting fatigue. As other teams started to get worn down as the season progressed, the Eagles won 7 of their last 8 games of the regular season, almost seeming to get stronger as the season went along.

Michael Vick, starting quarterback of the Philadelphia Eagles in 2013 in Kelly's first season, said there were noticeable effects on his overall well-being.

"I haven't felt like this in a long time . . . Come game time I just feel fresh and feel like I've been in my best shape in a long time."

Even though Chip Kelly only remained head coach for 3 years, his methods proved to be fruitful in his few years with the Eagles.

After leaving the Eagles, Kelly was hired as the head coach of the San Francisco 49ers. While he wasn't as successful as he was with the Eagles, the 49ers injuries were down again. As other teams begin to adopt Kelly's training regimen, I suspect we will see fewer soft tissue injuries. But in preventing injuries on the field, this technology can also help players improve their on-field performance as well.

Catapult is a major player in the wearable tech industry and should prove to be successful as their market begins to increase.

* * *

Artificial intelligence already plays a factor in the way athletes train through simulations.

A Silicon Valley based VR startup called STRIVR is working with NFL and college football teams to train their athletes virtually. The athletes don a headset in order to help them experience game scenarios without physically risking the safety and health of their bodies.

When athletes may be coming off an injury, they can still mentally train their mind and body with these virtual game scenarios. They may be limited physically, which keeps them out of practice, but now they can get a half-physical rep and a 100% mental rep.

STRIVR boasts clients such as the Dallas Cowboys, the San Francisco 49ers, and the Clemson Tigers. While this training can be useful, it is also expensive, costing upwards of six figures.

The company is also considering a transition into the consumer market, a reach further than they originally thought. STRIVR has created a goalie simulator, a chance for fans to put on a headset and stand in-net while hockey pros pepper them with virtual slap shots.

Virtual reality and simulation training is a reality in our sports world. Artificial intelligence has already worked its way into improving the way our athletes train.

These training simulations have more of an effect than one might think on the way athletes perform

* * *

Let's move on to AI and injury prevention.

When I was a sophomore in high school, I played on the JV football team. My mom was always worried about me getting injured. So much so that her nickname is "Safety Anne."

Nearly every play, she would scream if I got anywhere near the middle of the action. On this fateful day in the fall of 2012, I ran a seam route up the middle as a tight end. I caught the ball and I got tackled and wrapped up. I was being brought to the ground when the safety lined me up. One bone crushing hit later, the crowd gasped.

I laid on the ground for a few seconds (according the footage, I don't remember the incident).

The head athletic trainer brought me off to the side. *"Are you ok? You took a huge hit out there. Do you know where you are?"*

"Yea, I'm all good. We're at (my high school) and we're playing (the team we are playing)," I replied. And with that I was cleared to play.

Just one play later, my trainer brought me off to the side again. "You looked confused out there. Are you sure you're ok? Do you know where you are?"

But I didn't know. I didn't know where I was, or who we were playing, or why I was wearing pads.

"Did I catch it? Did I score a touchdown?"

These are the questions I repeatedly asked my Dad on the way home from the game. I couldn't remember asking these questions and yet I simply kept asking them.

Here is where wearable technology could come in handy. The

technology, based on previous athletes with concussions, could tell the trainers whether or not I had suffered a concussion.

Q30 Innovations has created a product to do just that. They have created a product with sensors to put on football athletes during a game. This effort is intended to gather data in order to determine what types of collisions cause concussions. While still in the early stages of its development, these sensors could eventually be a wearable technology answer for detecting concussions. Therefore, we can reduce the risk of exacerbating a brain injury further. I could've used this when I got my concussion; it would've saved me from returning to play.

* * *

Catapult Sports, a wearable tech company, also has their own developments in preventing injuries.

Have you ever noticed your favorite sports athletes wearing what looks like a sports bra over their training uniform? Well, that is a product of Catapult, a world leader in sports performance analytics. Basically this "bra" has GPS technology embedded inside of it that sends physical and positional data to the iOS app on a user's phone. Catapult is one of the first of its kind, deemed wearable technology. Wearable technology relies heavily on the use of GPS tracking systems that are able to collect tons of data sent to Catapult's app for analyzing.

Catapult and others are becoming increasingly more popular.

Catapult is well established and has clients that include professional soccer, basketball, and football teams. In an interview with City A.M., Benoit Simeray, Chief Executive of Consumer Products for Catapult, talking about his company's artificial intelligence capabilities speaks about it not only helping coaches to make decisions but also the possibility of being more effective than coaches themselves.

One of their biggest selling points is their GPS, but they are also focused on injury prevention. They have fine-tuned their algorithm for the avoidance of serious arm injuries for pitchers.

An interesting article about Catapult explains how an experiment was done with pitchers from a Major League Baseball team. They were asked to throw a quick bullpen session while wearing the Catapult baseball product. Instructed to take it easy, the pitchers each threw 25 pitches off the mound in order to get data to analyze their exertion from the workout. The data was crunched and a number of pitchers had exceeded their designated pitch allotment based on physical exertion on the elbow.

Data is gathered using an accelerometer, gyroscope, and magnetometer to measure the lean, the turn, and the force of the body. Catapult's team has developed a machine learning algorithm to recognize patterns of exertion and identify throws, swings, and sprints.

It is interesting to note the increase in elbow injuries in MLB. Over 400 Tommy John surgeries, a surgery to repair the UCL in the elbow, have been done from 1974 to 2015 on MLB pitchers. One third of these surgeries occurred from 2010-2015, a sharp increase. Catapult is attempting to gain data so that we may one day be able to limit these commonplace injuries.

Ryan Warkins, Catapult's director of business operations, believes the company's future is in sports specific implementations of their product, such as the one described above. A necessity for refining their algorithms is data. This is a recurring theme I have noticed in the analytics of sports. The only way these machine learning products will improve is through the collection of more and more raw data. While this can be sometimes tough to obtain, it is the all-important factor in bettering their product. Warkins acknowledges Catapult is still in the data collection and discovery phase but plans to use their mounds of data to create these sport specific products.

Warkins says that while they will be designing their products for professional markets, they eventually will target consumer markets. This coincides with Simeray's explanation of their soccer "bra" making the transition into the amateur market.

This transition into the consumer market really could pay dividends for companies like Catapult. The Little League World Series (LLWS) has instituted a pitch count system that indicates the number of days' rest needed after a certain number of pitches. Catapult could possibly complement their technologies to assist the LLWS in making their pitch count system more accurate than ever thought possible.

Simeray mentions the possibility that the technology could develop into being more effective than coaches themselves in making decisions for the team. This notion brings up a dangerous area of AI. Could coaches actually be phased out of their positions? Personally, I do not think so. Artificial Intelligence has proven to be advanced, but we are still struggling to replicate human behavior with AI. Human behavior is one the most complex models to create and mimic using technology. Certain aspects of coaching decisions, like knowing how a player is feeling on a certain day, are difficult, if not impossible, to replicate. It will be interesting to follow Catapult's transition into the consumer market, and I wish them the best of luck.

* * *

With technologies like these, it is easy to see the benefits. We can improve training, recovery, and reduce injuries.

Instead of having one training regimen for all the athletes, we can now customize routines for specific athletes. Every athlete handles injuries and recovery differently, so why treat them as if they don't?

We can even have a personalized weight training coach to maximize muscle growth. The applications of wearable technology are growing by the minute. They have proven to be successful before, and I'm sure they will again.

CHAPTER 8

FAN ENGAGEMENT

———

If you don't know who Steve Ballmer is, look up on YouTube something to the tune of "Steve Ballmer dancing" . . . I'll wait.

Yes, that is Steve Ballmer, the eccentric, billionaire owner of the Los Angeles Clippers. While on the surface he looks like any other goofy fan enjoying a Madonna song courtside at a basketball game, he is heavily invested in making the intersection of Artificial Intelligence and the NBA a reality. This does not come as a surprise to most, as Ballmer is the former CEO of Microsoft

An advisor to Second Spectrum, a company focused on using computer vision to track player and ball movement during a basketball game, Steve Ballmer helps guide the company's involvement with the NBA. On the verge of landing a contract with the NBA, computer vision could be the next best thing in sports entertainment.

* * *

Steve Ballmer has a large role in the Artificial Intelligence market. His intersection of technology and sports has fostered a desire to integrate AI technologies into the NBA in an effort to revolutionize the entertainment value of the NBA. Ballmer's association with Second Spectrum aims to provide tracking data to fans mid-broadcast. It is not hard to imagine Second Spectrum's entrance into the spotlight in a deal with the NBA.

"I think there's such an opportunity to change the way people experience sports by using technology."—Steve Ballmer, Los Angeles Clippers owner

In fact, the Clippers already will partner with Second Spectrum to release a new product that adds to the traditional TV broadcast with new data and animations driven by AI using computer vision. Viewers are able to watch and track player and ball movement at a granular level. Then with this information, it uses machine learning to derive new information for coaches, players, and beyond.

The technology lets fans see different plays develop, fantasy statistics, and even regions on the floor where a player might shoot from. Fans will have access to so many real time statistics that it could change the way we watch sports forever.

Its founder, Rajiv Maheswaran, was a former artificial intelligence professor at the University of Southern California. He was the man mentioned before who gave the TED talk on "moving dots" and how it enables predictive analysis on player movement on the court.

"We realized there was a lot of data coming into sports, and no one knew what to do with it."—Rajiv Maheswaran

Maheswaran's innovative thinking has put Second Spectrum seemingly ahead of their competition and into a deal with the Clippers. Going beyond just the Clippers, Second Spectrum's beta test spanned the 2017 NBA season and will be fully rolled out in the 2018 season for all NBA teams. It is not unreasonable to think that it could be integrated into the fan experience slowly over the next decade.

In an article published in 2016, Bloomberg reported that the NBA is about to enter into a 6-year, $250 million deal with Second Spectrum and Sportsradar, two companies discussed in later chapters. The deal would be to support the league's big data ambitions.

Sportsradar has already established itself in the American

professional sports market, signing deals with the NFL, NHL, and NASCAR. CNN even used Sportsradar for its Olympic coverage in 2016. Having added the NBA as a client, Sportsradar is poised for success, and so are its backers. It is funded by NBA owners like Mark Cuban, Michael Jordan, and Ted Leonsis.

Currently, STATS, a sports data analytics company, has a deal with the NBA to provide real time player data, but the impending move to Sportsradar would signal a big shift in the market. A large part of the deal centers around sports betting as Sportsradar also helps to provide real time data to bookmakers. Sports gambling is currently illegal outside of Las Vegas, but the NBA Commissioner Adam Silver has been a big proponent of legalizing sports betting across the nation.

The deal would combine Second Spectrum and Sportsradar technologies to help examine plays in more depth, by measures such as how they were defended or the likelihood of success. This technology would be an addition to the regular broadcast, therefore not putting media giants like ESPN and the big networks at risk, yet.

This deal signals a huge step forward for AI in the professional sports world. Multiple owners are investing in AI technologies to incorporate them into the NBA viewing experience. Obviously, with some modifications it can be applied to most professional sports markets and could potentially change the viewing experience of fans forever.

Fans drive the profitability of the professional sports industry. Without fans, professional sports wouldn't exist in the capacity it does today.

On a satirical sports podcast, PFT Commenter, one of two hosts, repeatedly jokes that the "fans pay the players' salaries." While obviously this is a satirical "hot take," one cannot argue that there is an element of truth to the statement.

Professional organizations are not only concerned about the success of their team, but when it comes down to it, their profitability and revenues. Buying into technology that only serves to provide the fans a better experience, which drives TV ratings, attendance, and ticket sales should be a serious consideration going forward.

* * *

Sportradar is a premiere supplier of sports data like STATS, and is preparing for a transition into the consumer market with its additional AI capabilities after its acquisition of MOCAP Analytics. MOCAP is a motion capture company that specializes in transforming granular data into information the average sports fan can enjoy to enhance their sports experience. Sportradar's mission is to supply player tracking and sports data to media companies, bookmakers, sports federations and government authorities. They also have fan-focused products that similarly supply fans with advanced data to enhance their viewing experience. Its acquisition of MOCAP will provide fans with reliable, advanced data at the tip of their fingers.

MOCAP's AI technologies have been used by sports entertainment corporations for years such as the NBA's Golden State Warriors.

According to the assistant GM of the Warriors, Kirk Lacob, they are *"the first NBA team to become heavily ingrained with player-tracking systems and companies such as MOCAP."*

Part of what made Sportradar's acquisition of MOCAP attractive to Sportradar was MOCAP's ability to transform such high-level, advanced data into information easily digestible for your average fan. They are attempting to establish a benchmark for access to this data for every fan across the world. Their partnerships with professional sports organizations has taught them that there exists a huge potential to raise fan experience by bringing out insights from the collected granular data. Their investment in MOCAP's machine learning approach shows their belief in the power of AI to enable their mission.

Sportradar currently has a product called Game Stream that is a real time player tracking tool for NFL games and is available to fans. Using 3D play visualizations built with NFL player tracking data, Sportradar provides an interactive visualization of the league's biggest plays in real time. With three different camera angles, fans can select the camera that they want and slow the play down as much as they want. The players show up as dots on the screen, and move at the speed at which the real players do, enabling the fan to see the play develop and the strategy behind both sides of the ball. Stats available through this product include Player Speed and QB throw time etc.

The acquisition of MOCAP brings along their similar technology called the "Machine." The Machine similarly displays dots on the field and displays stats for fan digestion. But it also adds another layer with AI. Machine learning and the player-tracking capabilities of The Machine allow for the analysis and prediction of plays before they even happen. Herein lies the motivation for the acquisition of MOCAP and their AI capabilities.

If this idea seems far-fetched, a recent interview with the CEO of Sportradar, Carston Koerl assures us that it is far from unrealistic; in fact, there is a market for it.

"It all sounds a bit like fantasy but the CPU power is there, and the video and data information is there; it just needs to come

together and gain traction."

A running theme I have noticed is companies' making the transition to the consumer market and allowing AI technologies to benefit the fan experience as well. Now fans will be able to access these stats with ease and impress their friends with the intimate knowledge of the game, forever changing the way fans engage in sports.

A possible future application of Sportradar's technology is its effect on the sports gambling industry. While currently sports betting is only legal in Las Vegas, Nevada, in the United States, there is the possibility for the influx of new statistics for fans to affect their betting strategies if legalized. While this may seem far off, there is current legislation that could eventually legalize sports betting in America. The outreach of such a product could not only have lasting effects on the way fans engage in viewing sports, but also the way in which we bet on the sports we love. Sportradar's acquisition of MOCAP's AI to aid their fan-centric products puts the company in position to become a major player in the lives of sports fans around the globe.

* * *

Fan engagement seems to drive the sports industry forward. But over the past few years, stadium attendance has decreased and so has television viewership. Specifically in the NFL, stadium attendance has dropped a concerning amount for the league that owns Sundays.

In an effort to combat this decline, some organizations are taking drastic steps to keep fans interested over the course of a game. Sometimes the halftime celebrations just don't cut it for the newer generation. That said, the Red Panda halftime performance is one for the ages; it is truly exhilarating.

Certain companies and organizations are employing AI technology to create chatbots that engage with fans. These AI bots generally are able to answer fans' questions not only about anything in the game, but also about the locations of hot dogs stands within the stadium. This is done through a process of natural language processing. Essentially the bot is given an understanding of what it needs to answer, but after more and more conversations, the machine learning algorithms are trained to become more accurate.

For example, the Sacramento Kings, an NBA team, is one of these professional teams on the cutting edge of Artificial Intelligence. The team recently introduced to fans their bot, aptly named Kai, short for Kings Artificial Intelligence.

"The team's new bot will be able to support fan inquiries about the Kings and Golden 1 Center, and will continue to learn with

each interaction with the loudest fans in the NBA.

On Friday, the Sacramento Kings in partnership with JiffyBots, introduced its Facebook Messenger bot.

Kai, which is an acronym for Kings Artificial Intelligence, will be able to answer a wide range of questions, such as info about Sacramento's roster, stats, Golden 1 Center details, franchise history and more.

In addition, Kai will present users with a unique one-on-one chat experience that will evolve with each correspondence."

The best part for Kai is that, the more conversations it has, it will make fewer mistakes and give more accurate answers.

Another chatbot that operates in a similar fashion comes from our friends across the pond at Wimbledon. Their bot is named Ask Fred, a product of IBM. By using a natural language interface, it understand what visitors are looking for and is ready to serve them. By leveraging real time information, Ask Fred's AI can also predict winning players/teams, predict team dynamics, and familiarize venue operations, creating the best possible fan experience. The efficiency of Ask Fred's answers sustain the engagement of Wimbledon fans.

Another leader in the technology industry, Microsoft, also has invested in its own version of chatbot, the Bing Sportscaster Bot. Its mission is to provide fans with the latest sports news. Covering the majority of professional sports organizations, it can give fans roster information, game fixtures, results, in-game information, and even predict the winner of games.

Also from our friends at Sky Sports across the pond, soccer fans can engage with Jeff Bot, a Facebook Messenger bot intended to mimic the Sky Sports Soccer Saturday host Jeff Stelling. Providing similar information as the Bing Sportscaster bot, Jeff Bot drew 10,000 sign ups within 24 hours of its launch. It's aimed at keeping fans engaged with the conversation about their favorite Premier League and Champions League sides.

"*People are treating Jeff Bot like a person and that is step one in creating something powerful.*"—Alex Beckman, Founder and CEO of Gameon, the creator of Jeff Bot

Successful chatbots, like the examples above, seem to contain a great propensity for human tendencies. On top of intelligence, responsiveness, and entertainment, personality and identity seem to be distinguishing characteristics that the good conversational bots have versus the failed conversational bots.

The importance of treating these bots as if there were another human on the line results from its natural language processing algorithms. The only way to improve their ability to understand the human language is to be submerged in it on a daily basis.

"AI and humans perform best when they work together and can trust each other."—Big High, CTO at IBM Watson

Customer engagement is being redefined with the power of AI and cognitive computing, especially in the sports industry.

CHAPTER 9

BEING A FAN

——

As we start to get away from technical definitions of Artificial Intelligence, the human element gets increasingly more prevalent. Artificial intelligence can be framed as an attempt to take away jobs that humans could otherwise occupy. This certainly will be the case for some people, but in most cases AI technology aims to assist humans in their endeavors.

"Machines will be capable, within 20 years, of doing any work a man can do."—Herbert Simon (1965)

While Mr. Simon here was significantly off, his hypothesis might prove true in the coming years. Yet, there is still a long way to go before this is the case. Is AI just a tool that humans use to assist in daily tasks? Or does AI have greater aspirations for replacing humans in their jobs?

For example, the Associated Press, a large newspaper company with ample resources, is not able to send sports journalists to every Minor League Baseball game. There are over 200 minor league teams. Even a giant like AP cannot cover all these games, as they just don't have the numbers to do so. Or do they?

* * *

As we have already seen, the applications of Artificial Intelligence are endless. No matter the field or division of sports, AI seemingly will intervene in some potentially positive way. While most of these products create a positive impact on the sports industry, some may have lasting impacts on employment in the future. For example, the Associated Press is using AI technology to write sports articles about minor league games, calling into question the job security of MiLB journalists.

This AI "reporter" is able not only to read the data and return facts from the game, but also to analyze the data, pull highlights of the game, and construct a well-written article, given this detailed information. They are in a partnership with MLB Advanced Media, who provides them with the detailed statistics from the game as they are the official statistician of Minor League Baseball. Here is an excerpt from an article written with this technology.

"Cristian Alvarado tossed a one-hit shutout and Yermin Mercedes homered and had two hits, driving in two, as the Delmarva Shorebirds topped the Greensboro Grasshoppers 6-0 in the second game of a doubleheader on Wednesday.

Alvarado (6-4) struck out eight and walked one to pick up the win.

In the bottom of the first, Delmarva took the lead on a solo home run by Mercedes. The Shorebirds then added four runs in the third and a run in the fourth. In the third, Steve Laurino hit a two-run single, while Ricardo Andujar hit a solo home run in the fourth."

While this article clearly reads coherently and is well-written, it does lack the flair that a sports journalist would add to it. This simply reads as a good summary of the game rather than personalized reporting.

AP's deputy director of sports products Barry Bedlan says that there will initially be human editors involved, but a need for human editors will be eliminated as the system begins to become more accurate and intelligent. This requires extensive training on the data, because the AI is pointless if some part of the article it creates is incorrect.

While it is still in the early workings of its development, it can already produce stories that sounds as if they could have been written by a human, albeit a boring human. Some feedback that AP has received revolves around the human component of journalism. This "reporter" is currently lacking the "colorful language" that journalists have been known to use in their articles.

Another advantage to the AI reporter is the wide range of stories and articles it could write with ease. Normally, a journalist will need time to create a well-written and thought-out article about a game that they took time to watch and study. The AI can read the statistics from after the game, compose a story, and publish in less time than it would take a sports journalist.

AI reporting is not a new frontier for AP. Before the MiLB, they used a similar technology to compose earnings reports in the financial sector. According to their News Automation Editor Justin Myers, they went from being able to follow and produce 400 quarterly earnings reports to about 4,000 just with this AI technology, a staggering nearly tenfold increase. With over 200 clubs in the Minor League system, it is not a far jump to say their technology could cover every team and every game.

One concern that I have is the lack of human connection to the stories. I believe the challenge for AP is to create stories that not only read like a normal article, but that add the colorful language usually present in a sports article. With earnings reports, it seems an easier transition from pulling information from a 10k form into a line by line summary of the earnings data. On such a report, the usual flair from journalists is not present. This is the biggest difference between the financial sector and the sports journalism industry. If AP is able to add that extra human component to its technology, I would start to worry for the job security of your favorite sports journalists. But for now, I would rather read an article with some pizzazz versus a purely factual summary of my favorite team's performance from the night before.

* * *

The Olympics are a gigantic stage where elite athletes gather from around the world to compete in their respective sports. At Rio in 2016, athletes competed in 28 total sports, each with preliminaries, quarterfinals, and finals. With so many races and events to keep track of, it is hard to cover them all. I know I can only keep up with a few sports that I enjoy and there's not room for much else.

This is how the Washington Post has used AI to keep their readers up to date with each event. Their Olympics twitter account used an AI software powered by the sports data company STATS. Their goal was to post about 600 Twitter updates and 300 short narrative articles revolving mostly around raw data using this software, endearingly named Heliograf (@ WPOlympicsbot). The intent was for Heliograf to post 1 to 3 sentence updates on things like current medal counts.

While it is an amazing product, the Washington Post says that Heliograf will not have any effect on journalists' jobs. In fact, they've hired over 140 journalists since 2013. The goal rather is to assist the journalists in covering more of the individual events. They want humans to find the sources, discover the stories, and provide the analysis that their readers want to envelop themselves in.

Another assurance the Post made is that every story published by Heliograf will be seen by a human editor. All stories published will have a human insuring that nothing Heliograf says or does has errors. The Post shows more trust in their journalists and less trust in AI than the Associated Press, which has started to have similar AI bots covering Minor League Baseball games.

This discrepancy in trust in AI is expected when new technologies come about. My belief is that, no matter what level of trust you have in your journalists or your AI, humans must be involved in order to get a successful product. Now this may change as AI bots get more advanced in imitating human writing styles. But for now, the Post's Heliograf bot needs a considerable amount of human oversight to function properly.

In an age when the ability to create new, interesting content is necessary, if these bots are going to help provide fans with more content, then I think the bots will be incredibly helpful in doing so.

* * *

In my senior of high school, I was playing for the high school baseball varsity team. We were up against our rival Calvert Hall. The score was 5-4: we were up one run in the second to last inning. While playing second base, I made an error on a ground ball hit to me. In the biggest game of the year, I made a huge mistake that let two runs in. It just so happened that the lacrosse team got out of practice before the end of our game and most of them stayed to watch on the sidelines.

But luckily, I was given an opportunity to redeem myself. I found myself facing Calvert Hall's best pitcher with runners on second and third base. We were down 6-5 in the last inning. Going into the game, my swing was feeling good, so I was confident. 3 pitches later I was in a 1-2 count with everyone watching. I reached for a curve ball and slapped it towards shortstop. I have to admit, it wasn't hit well. The shortstop fielded it on the run to his right. It ended up being a tough play for the shortstop, because I didn't hit it hard enough.

"There's no way. Just no way."

As luck would have it, the shortstop threw it away and 2 runs scored, giving us the win. I couldn't believe it. I got to first base as the winning run crossed the plate to everyone's glee. At that point, I was only kind of happy with myself, given the perfectionist that I am. I was pretty disappointed that it was not a hit but only a walk-off error. I'll remember that moment for as long as I live my life. Although it wasn't a hit, I don't really care, because at that moment I was a hero if only for a day.

Although this memory will exist in my mind forever, an article published on it could have made it more noteworthy outside of our small high school community.

Over the past couple years, I have assisted in coaching my younger brother's travel baseball team. I love it because I get to develop a special mentoring relationship with kids on the team.

When I have to miss the game, the group of 14 year olds tease me, *"Ooooooh are you hanging with your GIRLFRIEND??"*

But even if I am with friends or at work, I am able to follow the team on the Gamechanger scoring mobile application.

Instead of using pen and paper as in the old days, Gamechanger presents a scorekeeper digital version of scoring. Fans, like myself, can get real time updates on the current game and see past stats.

But Gamechanger recently teamed up with Narrative Science to provide a premier product for those who cannot attend or even follow the game. It has a new feature called Recap Stories, which uses the data captured from the game to craft a newspaper-worthy article covering the highlights of the game.

How does it work?

Narrative Science uses Advanced Natural Language Generation, a subset of artificial intelligence to create "intelligent narratives." Specifically, they describe Advanced Natural Language Generation as . . .

"A subset of Artificial Intelligence (AI), Natural Language Generation (NLG) creates automated, data-driven communications. In its simplest form, NLG turns structured data into text. But Advanced NLG, the AI software behind Quill, uses your intent as its guide to deliver the information you care most about. Applying advanced analytics to assess the data, it automatically transforms the resulting insights into human-sounding narratives at scale."

Compared to Associated Press's and The Washington Post's similar products, this serves as the least threatening to the journalism profession. It offers a recap of amateur games that would not be covered otherwise. Its introduction into sports at the amateur level could work its way up to the professional level, making it a competitor to AP or the Post.

Narrative Science, the provider of this technology, has 30 similar clients that employ their technologies to write various articles, covering niche fields like sports, finance, and real estate. What makes Narrative Science attractive to their clients is that they are low-cost and instantaneous. A sports journalist might be paid to write an article in an hour or two, while the AI technology can produce a story in under a second.

In a recent article in Slate, Evgeny Morozov summarizes the potential implications of automated journalism.

"Given all this, the idea that greater automation could save journalism seems short-sighted. However, innovators like Narrative Science are not to blame; used narrowly, their technologies may actually save costs and perhaps even allow some journalists— provided they can keep their jobs!—to pursue more interesting analytical projects rather than rewrite the same story every week."

What it comes down to is the extent to which this technology is applied.

In the case of the Associated Press, they are more liberal in their application of AI-based automated journalism. They are producing articles for nearly every minor league baseball game in seconds, creating potential problems for current sports journalists.

The Washington Post was clearly more hesitant to widely deploy their automated journalists. They made a point of clarifying that it will not endanger the jobs of sports journalists. It is meant to aid the sports journalists, potentially allowing them to pursue other, more diverse articles.

In the near future, automated journalism technologies are meant as an aid for journalists. But looking towards the more distant future, sports journalists might want to keep their career options open.

* * *

Continuing the story about the Associated Press's coverage of 142 MLB affiliated minor league baseball teams, Danny Page, a writer for Medium, weighs in on its lasting effects on the sports journalism industry. He claims that the future of sports journalism will not be automated.

"Done properly, automated journalism has the potential to make all our jobs more interesting."—John Micklethwait, Bloomberg News editor-in-chief

As I said above, it frees up journalists to investigate better stories than a Minor League Baseball game. If a journalist were to write an article for a MiLB team, the article would contain little more than the box score anyways, right?

Well, that is isn't always the case. For example, the Sonoma Stompers started two women in the game, but the automated recap might only include how Kelsie Whitmore hit for a double in the bottom of the third. The automated recap lacks the understanding needed for capturing this historic moment.

The stories seem to be missing a few things. They are short and impersonal as if Siri had written them. But this doesn't have to be the case. Yahoo has been using Wordsmith from Automated Insights to offer customized weekly fantasy-football recaps. After a request for a snarkier tone, it was found that clichés are programmed easily, a possible step towards making a more personable story.

Now, to look at a women's professional soccer game featuring the Washington Spirit vs. the Orlando Pride, the game coming days after the deadly shooting of 49 people in a club in Orlando. The automated recap might simply note that Orlando held an extra minute of possession from the 48th to the 49th minute, missing the gravity of the situation. Between the 48th and the 49th minute, the players and fans stopped to pay tribute to those killed in the deadly shooting. The emotion and the human element of a moment like this needs to be captured by a human journalist able to understand the context of the game, not just analyze the data.

Every time I watch the video of the time stoppage, a tear comes to my eye due to the extreme gravity of the moment. My thoughts transcend the game entirely, thinking of those we lost and their families. Taking this type of emotion out of journalism would be a detriment to the gravity of sports and what they are able to represent.

Take, for another example, the Miami Marlins game on September 26, 2016. José Fernández, a young, rising star in the Marlins organization, tragically died the day before on the 25th. After a pregame ceremony in his honor, Dee Gordon, the starting Marlins second baseman, steps up to the plate for the Marlins first at bat of the game.

Gordon, a lefty hitter, stands at the plate in the right hand batters' box, a tribute to his late friend Fernández. After taking the first pitch for a ball, Gordon switches over to his normal side. Just one pitch later Gordon blasts his first home run of the year over the right field fence. The crowd goes nuts. Gordon rounds the bases, points to the sky, touches home plate and breaks down crying. Even writing this gives me the chills.

Similar to the women's soccer game in Orlando, the gravity of this play transcends statistics and sports. Without human writers, the appreciation of this moment would go under the radar, simply documented as a home run.

I believe that the automated journalist industry is profitable, but limited. There is inherent value in distilling numbers and making them more readable for humans to interpret. But if we are unable to instill the human element and understanding into an AI driven journalist, then I fear for the sports journalism industry.

Many journalists are able to express the human side of the game. These are the stories not being covered by the AP or the Washington Post. And these are stories to be told; it would be a shame if they weren't.

Sports journalists will become expendable if they do not adapt the way that they write their stories.

* * *

Beyond reading about sports, the way in which we watch sports is constantly changing. And technology is what drives this change.

I have talked about how computer vision technology continues to advance. Out of this comes FreeD from First Vision. FreeD gives the fan a chance to experience the game from a player's perspective. Using an integrated camera system attached to a player's uniform, fans can now watch the game from the perspective of their favorite player.

Using 3D and virtual reality technology, FreeD gives a 360 degree perspective from the position of, say, Lionel Messi. In order to do so, 28 high definition cameras are placed around the stadium to capture video. First Vision's technology then compiles all the views into one seamless camera view.

The personalization of one's view is a revolutionary idea. Instead of the network's cameraman dictating what you see at every turn of the game, the fans can select the view that they please. It is almost as if they were a player sitting on the bench, getting a 360 degree view of the plays on the field or court.

FreeD is already being used for TV broadcasts in the NBA.

As technology gets more advanced, our viewing seems to get more personalized and realistic.

"Technology will not only change amateur and professional sports, but also the way we as spectators experience sport in front of our screens or in the stadiums."—Medical Futurist

CHAPTER 10

WOULD WE EVER REPLACE THE HUMAN ATHLETES?

———

The one thing we haven't *really* discussed is whether computers could actually replace the humans involved in sports? I mean why would anyone actually *watch* a sport involving a machine anyways? What fun is that? Wouldn't that just be like watching a video game play *itself*?

But what about a sport like auto racing?

Few people know of the advancements in self-driving cars that were made even 20 years ago.

While the rise of media coverage has been great for the world of AI, it has also forgotten the progress that many talented people have made over the years. A popular story of AI is the self-driving car. Tesla and Google are competing with each other trying to create a self-driving car. While these may seem like new technologies, they are actually just new and improved technologies. Self-driving cars have existed for two decades, but these stories did not see the light of day to the general public. Stanley, a self-driving car, drove nearly 150 miles across the desert in 2005 on a human made course.

Granted, Stanley wasn't on the highway, potentially putting human lives at risk. But these are the advancements that have been made over the years. AI growth has resulted from better computing power, refined algorithms, and more attention. It will only continue to grow upwards from here.

* * *

Driverless car applications can help so many different industries. But how can it affect sports? Can we replace human drivers?

Driverless cars are beginning to make their way into the sport of racing. We could be on our way to seeing drivers being phased out of racing.

For example, DevBot is a driverless prototype made by London-based Kinetik for the upcoming Roborace championship. The company has been testing its driverless car at circuits around the world since August 2017.

While DevBot still has a cockpit that allows a human to maneuver it around the track, it also has the capability to drive and race all on its own.

In a separate model Roborace Robocar, the cockpit has been eliminated. Now with no human influence, only artificial intelligence, its aim is to compete on a real race circuit. It almost resembles a Tron bike in its beauty and simplicity.

The Robocar could eliminate the human error in racing, a main reason that makes racing so exciting. Where is the fun in a car that doesn't feel the pressure and the fear of crashing at every turn?

Robocar could prove to make up for its lack of human error. One benefit is faster speeds. Cars would no longer be inhibited as much by G-forces and crashes, and drivers unwilling to reach top speeds in order to avoid a crash.

Similarly, it greatly reduces the possibility of injury or death. Even without humans, though, there is a possibility for crashes, so viewers don't lose all the thrill of a crash. Two Roboracers faced off at the Buenos Aires Formula ePrix, and one crashed into the barriers to avoid a dog on the track. The artificial intelligence is so precise that it recognized a dog in the road and swerved to avoid it.

But garnering excitement and viewership from fans isn't the only reason Kinetik developed these racers. They were developed in part due to the vision of Denis Sverdlov, founder of Kinetik and Roborace.

"We passionately believe that, in the future, all of the world's vehicles will be assisted by AI and powered by electricity, thus improving the environment and road safety . . . It's a global platform to show that robotic technologies and AI can co-exist with us in real life."—Denis Sverdlov

Driverless cars obviously have other applications that can help day-to-day life, but Kinetik started with their technology in the motor racing circuit.

AI in the form of driverless cars are already making an impression on the racing circuit, and are performing just as well, if not better, than human drivers.

Engineers at Stanford University have shown that their souped-up Audi TTS named Shelley can now outperform skilled racing drivers. In 2015, they pitted Shelley against the track expert at Thunderhill Raceway Park in Northern California. Shelley beat the expert by 0.4 of a second.

The engineers designed the car by meticulously studying human drivers, going as far as attaching electrodes to their heads to monitor brain activity. They hoped to learn which neural circuits are working during difficult maneuvers.

Using these results, they created Shelley, which can drive better than expert drivers.

"We've been trying to develop cars that perform like the very best human drivers"—Professor Chris Gerdes, director of the Revs Program at the Center for Automotive research at Stanford University.

What they found through their research is, that when drivers execute difficult turns, they actually use less brain power, relying more on muscle memory and instinct.

"If you are thinking, you are going too slow."—David Vodden, CEO and the expert at Thunderhill Raceway Park

Instead of relying on human power to maneuver around turns, Shelley uses a stabilizing algorithm that acts as if it were muscle memory. If Shelley started to slide, the algorithm would correct the slide with a motion similar to that of an expert driver.

Shelley isn't ready to go out and beat the best of the best just yet. It still has some work to be done in order to compete with the best racing drivers.

"In the future it could mean we can make driverless cars which drive as well as the best racing drivers. You could have a car with the skill of Michael Schumacher taking your kids to school or the dentist."—Joe Funke, PhD student at Stanford Revs Program

The advantage of Shelley lies in its rule following. Human drivers were found to break their own rules when it comes to driving their vehicle. Shelley, on the other hand, follows its rules for driving as programmed, which signals its full potential to outrace human drivers.

* * *

AI and virtual reality applications have intersected with auto racing in an interesting way. Jann Mardenborough's story illustrates this connection. Jann was always a quiet kid growing up, but always knew that his one love was for racing.

But reality kicked in, and at the age of 11, Jann had to stop racing due to the financial burdens on his family. Returning to his bedroom, Jann remained reserved as he imagined what could've been.

But throughout, Jann never lost his love for racing. So he turned to the video game Gran Turismo, a game in which one races virtual cars on renderings of real life tracks. Gran Turismo is a little more involved than just pressing some buttons on a controller, though. It is less a video game and more a simulator of real racing, where one sits in a chair with pedals, a steering wheel, and a screen. After hours and hours of practice, and drawing on his previous racing experience, Jann qualified for a Gran Turismo online competition.

After placing 8th in the competition out of 90,000 participants, Jann won the chance to actually drive a real car against other winners at Brands Hatch, a motor racing circuit in England.

After stopping racing at 11 years old, Mardenborough sat at the wheel of a serious racing car in the Dubai 24 Hour race. From his bedroom to the race track, he made himself into a competitive and successful race car driver.

Even now, current Formula One drivers use Gran Turismo to prepare for certain tracks and races. Lewis Hamilton, one of the best Formula One Racers in the world, admitted to using Gran Turismo during his rookie year.

The game is hyper realistic when it comes to mimicking reality.

In cockpit mode, it takes little conscious effort to suspend belief that you are in a real race car.

There is one big difference— the simulations still lacked movement. It is hard to mimic the sensation of the car reacting, the grip felt through the set, the G-forces that compress the body, and the forces generated when turning a corner. But still, Gran Turismo has proven to be a useful simulator for training racing professionals.

Sony, the creator of Gran Turismo, teamed up with Nissan to form the GT Academy in 2008. The project aimed to see if you could take a gamer and put them in a real racing car with success. In fact you could.

Lucas Ordeñez, a 23 year old from Spain, won both the online and real-world challenge. After some serious training, he was able to compete in the Dubai 24 in 2009, just one year after the academy was formed.

But Nissan was obviously worried that Ordoñez still was not good enough to compete, possibly putting his life in danger.

"I'm not a nervous guy, but I was physically sick with worry that we were sending this guy out to his death," said Nissan's Darren Cox.

Lucas changed their minds with the way that he calmly handled problems on the track. With a flat tire, he instructed the crew that he would be coming in for a pit stop to change a flat tire.

Nissan was then convinced that they might actually have a viable simulator. With two more racers joining professional racers in the next few years, one being Mardenborough, Nissan's GT Academy has been a success. Surpassing all doubters, the racers trained by the simulators seemed to be naturals on the racetrack.

"It felt completely normal . . . I'd never power-steered a car before, I had only ever done it in a game. I was controlling it just with the throttle and it was completely natural to me."—Mardenborough

Mardenborough not only became a professional racing driver through Gran Turismo, but also gained confidence and swagger that he never would've had otherwise. Nissan's GT Academy could pave the way for virtual training of athletes.

* * *

Training using technology and simulators has proven to be successful. Technology has also given us driverless cars that can compete with professionals. Both have proven to be successful aides to the sports we watch. But how intrusive is this technology into our sports?

In racing, we can see that driverless cars are already competing in their own races, a leap from Stanley's guided journey in 2005, but not a gigantic one. Driverless cars have slowly progressed to a point where they could compete in a real racing circuit. Now, we don't know if they'll stay separate from real race car drivers, but I doubt it. I think the integration is inevitable.

Is this viable in other sports? Could athletes ever be replaced by robots?

There are two important considerations, in my opinion, that will play a large role going forward on this subject.

The first is the advancements that AI has made in recent years. AI is a powerful tool that is still underutilized in our society today. It is clear that the potential to create these robots is there. Robots that compete at a professional level are certainly a possibility, although possibly far in the future.

The second is the human element. I truly believe the only thing stopping the robots from replacing our athletes is the humanity that is present in sports. Like the story of Jesse Owens, we as humans hang on the stories of perseverance and overcoming adversity. Without this, sports wouldn't be the great spectacle that we know and love.

This is what will keep the robots from replacing our athletes. This does not prevent us though from striving to create machines that play our sports. Motor racing is just the beginning of the integration of machines into professional sports. But I do not believe that athletes will be banished for good.

REFERENCES

___

Appolonia, A. (2017, October 20). Elon Musk's artificial intelligence
company created virtual robots that can sumo wrestle and
play soccer. Retrieved February 26, 2018, from http://www.
businessinsider.com/elon-musk-openai-virtual-robots-learn-
sumo-wrestle-soccer-sports-ai-tech-science-2017-10

Beckett, J. (2016, October 27). Hitting it Out of the Park with Deep
Learning | NVIDIA Blog. Retrieved February 26, 2018, from
https://blogs.nvidia.com/blog/2016/10/25/deep-learning-base-
ball-analytics-top-statcast/

Bishop, T. (2016, April 01). Interview: Paul Allen's artificial intelligence guru on the future of robots and humanity. Retrieved
February 26, 2018, from https://www.geekwire.com/2016/

geekwire-radio-paul-allens-artificial-intelligence-guru-fu-ture-robots-humanity/

Bolton, J., Professor. (2017, October 18). Computer Vision and Neural Networks [Personal interview].

Booton, J. (2017, December 26). Golden State Warriors Andre Iguodala On Wearables, Tech Investing, Golf. Retrieved February 26, 2018, from https://www.sporttechie.com/war-riors-andre-iguodala-talks-wearables-and-tech-investing/

Darrow, B. (2015, September 4). A Sneak Peak Into MLB Statcast Live (Sort of) From Fenway Park. Retrieved February 26, 2018, from http://fortune.com/2015/09/04/mlb-statcast-data/

De Jesus, C. (2016, July 07). AI Will Begin Writing News Stories for Minor League Baseball Games. Retrieved February 26, 2018, from https://futurism.com/ai-will-begin-writing-news-stories-for-minor-league-baseball-games/

Hall, J. (2017, June 17). Can Catapult Sports become the Strava of football? Retrieved February 26, 2018, from http://www.cityam.com/266852/can-catapult-sports-become-stra-va-football-interview | https://www.sporttechie.com/catapult-harnesses-ai-help-solve-baseballs-injury-problems/

Helling, D. (2017, October 12). Sportradar MOCAP acquisition brings

'The Machine' to sports fans. Retrieved February 26, 2018, from https://fansided.com/2017/10/12/sportradar-mocap-acquisition/

Kay, A. (2017, May 11). Kentucky Derby Partnering With A.I. Company That Correctly Predicted Last Years Superfecta. Retrieved February 26, 2018, from https://www.forbes.com/sites/alexkay/2017/05/03/kentucky-derby-partnering-with-a-i-company-that-correctly-predicted-last-years-superfecta/#188e15d-02aba

Korosec, K. (2017, February 27). Roboraces Electric Race Car Doesn't Need a Human Driver. Retrieved February 26, 2018, from http://fortune.com/2017/02/27/roborace-electric-robocar/

Le, H. M., Carr, P., Yue, Y., & Lucey, P. (2017, March 3). *Data Driven Ghosting using Deep Imitation Learning*[Scholarly project]. In *Disney Research*. Retrieved from https://www.disneyresearch.com/publication/data-driven-ghosting/

Liberatore , S. (2016, July 01). Your days could be numbered if you're a sports writer: The Associated Press is using AI to write Minor League Baseball articles. Retrieved February 26, 2018, from http://www.dailymail.co.uk/sciencetech/article-3668837/Your-days-numbered-sports-writer-Associated-Press-using-AI-write-Minor-League-Baseball-articles.html

Logothetis, P. (2017, November 04). Cracking the vault: Artificial intelligence judging comes to gymnastics. Retrieved February 26, 2018, from https://www.theguardian.com/sport/blog/2017/nov/04/ai-judges-gymnastics-olympics

Lucey, P., & Felsen, P. (2017, March 4). *"Body Shots": Analyzing Shooting Styles in the NBA using Body Pose*[Scholarly project]. In *MIT Sloan Sports Analytics Conference*. Retrieved from http://www.sloansportsconference.com/wp-content/uploads/2017/02/1690.pdf

Lucey, P. How AI-Based Sports Analytics Is Changing the Game. (2017). Retrieved February 26, 2018, from https://adtmag.com/blogs/dev-watch/2017/07/sports-analytics.aspx

Maffei, L. (2016, August 05). Robots will cover the Olympics for The Washington Post. Retrieved February 26, 2018, from https://techcrunch.com/2016/08/05/robots-will-cover-the-olympics-for-the-washington-post/

Maheswaran, R. (n.d.). The math behind basketball's wildest moves. Retrieved February 26, 2018, from https://www.ted.com/talks/rajiv_maheswaran_the_math_behind_basketball_s_wildest_moves/transcript#t-14043

Maloof, M, Professor. (2017, October 13). The Evolution of AI over the Years [Personal interview].

Maloof, M., Professor. (2016, August 26). *Artificial Intelligence: Your Phone is Smart, but Can it Think?* [Scholarly project]. Retrieved from http://people.cs.georgetown.edu/\~maloof/pubs/prelude16.pdf

Morris, B. (2016, December 14). Virtual-Reality Startup Strivr Raises $5 Million in Initial Funding Round. Retrieved February 26, 2018, from https://www.wsj.com/articles/virtual-reality-start-up-strivr-raises-5-million-in-initial-funding-round-1481726884

Neal, R. W. (2016, March 31). Wealthfront Turns to Artificial Intelligence to Improve Robo Advice. Retrieved February 26, 2018, from http://www.wealthmanagement.com/technology/wealthfront-turns-artificial-intelligence-improve-robo-advice

O'Conner, P. (2017, October 19). Insight from the President of MiLB [Telephone interview].

Richards, G. (2015, February 12). Jann Mardenborough reaches top with Nissan from virtual video game grid. Retrieved February 26, 2018, from https://www.theguardian.com/sport/2015/feb/12/jann-mardenborough-nissan-le-mans-24-playstation-world-endurancempionship

Roof, K. (2015, October 27). Michael Jordan, Mark Cuban, Ted Leonsis Betting $44 Million on Sportradars Data. Retrieved February 26, 2018, from https://techcrunch.com/2015/10/27/

michael-jordan-mark-cuban-betting-44-million-on-sportra-
dars-data/

Rosenberg, M. (2016, August 18). What is the state of Artificial
Intelligence in sports? Retrieved February 26, 2018, from https://
www.si.com/tech-media/2016/08/18/olympic-medalist-tech-
nologist-descibes-future-ai-sports

Ruiz, E. (n.d.). Game Recap Stories. Retrieved February 26, 2018,
from https://gamechanger.zendesk.com/hc/en-us/arti-
cles/212072186-Game-Recap-Stories

Sennaar, K. (2018, January 17). Artificial Intelligence in Sports—
Current and Future Applications. Retrieved February 26,
2018, from https://www.techemergence.com/artificial-intelli-
gence-in-sports/

Skweres, A. (2017, June 02). AI and the Growing Use of Technology in
Sport. Retrieved February 26, 2018, from https://www.stats.com/
industry-analysis-articles/ai-growing-use-technology-sport/

Soper, T. (2016, August 12). NBA reportedly nearing $250M deal with
sports data firms Sportradar and Second Spectrum. Retrieved
February 26, 2018, from https://www.geekwire.com/2016/
nba-reportedly-nearing-250m-deal-sports-data-firms-spor-
tradar-second-spectrum/

Sports Betting: The Next Big Thing for Artificial Intelligence. (2017, June 12). Retrieved February 26, 2018, from http://www.publicgaming.com/PGRI/index.php/news-categories/sports-betting-daily-fantasy-sports/1212-sports-betting-the-next-big-thing-for-artificial-intelligence

Taylor, T. (2016, December 27). The big change to wearable tech you'll see in 2017. Retrieved February 26, 2018, from https://www.si.com/edge/2016/12/27/sports-tech-2017-wearable-technology-future

The Man in the Arena. (2012, June 03). Retrieved February 26, 2018, from http://www.theodore-roosevelt.com/trsorbonnespeech.html

Vincent, J. (2017, July 06). This startup is building AI to bet on soccer games. Retrieved February 26, 2018, from https://www.theverge.com/2017/7/6/15923784/ai-predict-sport-betting-gambling-stratagem

Yue, Y., Carr, P., Lucey, P., Bialkowski, A., & Matthews, I. (n.d.). *Learning Fine-Grained Spatial Models for Dynamic Sports Play Prediction*[Scholarly project]. Retrieved from https://pdfs.semanticscholar.org/76a6/216dc18a703a5f5913e0bd0c5e5027d-cdc6f.pdf

ACKNOWLEDGEMENTS

———

I have had the unique opportunity to dive into the intersection of two fields that I love. I would like to recognize the people that helped me achieve my goals. Each conversation only furthered my love for sports and technology. Thank you to my interviewees who gave me a chance to learn more. Your time is appreciated, and I am forever grateful for your support.

Next, my family. Mom, Dad, Matt, and Uncle Stephen, thank you for your unending support. I wrote, even before I started the writing process, that it would be a proud moment if I were to hand my parents my finished book. Well, that moment has come, and I owe it to them. A special thanks to Mom and Uncle Stephen who lovingly poured over every word.

And for all those I missed, thank you for input and

wholehearted support of me. It means the world.

www.ingramcontent.com/pod-product-compliance
Lightning Source LLC
Chambersburg PA
CBHW071523180526
45171CB00002B/359